Lessons from Nanoelectronics
A New Perspective on Transport
— Part A: Basic Concepts

Lessons from Nanoscience: A Lecture Note Series

ISSN: 2301-3354

Series Editors: Mark Lundstrom and Supriyo Datta
(Purdue University, USA)

"Lessons from Nanoscience" aims to present new viewpoints that help understand, integrate, and apply recent developments in nanoscience while also using them to re-think old and familiar subjects. Some of these viewpoints may not yet be in final form, but we hope this series will provide a forum for them to evolve and develop into the textbooks of tomorrow that train and guide our students and young researchers as they turn nanoscience into nanotechnology. To help communicate across disciplines, the series aims to be accessible to anyone with a bachelor's degree in science or engineering.

More information on the series as well as additional resources for each volume can be found at: http://nanohub.org/topics/LessonsfromNanoscience ·

Published:

Lessons from Nanoscience:
A Lecture Note Series

Vol. 5

Second Edition

Lessons from Nanoelectronics

A New Perspective on Transport
— Part A: Basic Concepts

Supriyo Datta

Purdue University, USA

 **World Scientific**

NEW JERSEY · LONDON · SINGAPORE · BEIJING · SHANGHAI · HONG KONG · TAIPEI · CHENNAI · TOKYO

Published by

World Scientific Publishing Co. Pte. Ltd.
5 Toh Tuck Link, Singapore 596224
USA office: 27 Warren Street, Suite 401-402, Hackensack, NJ 07601
UK office: 57 Shelton Street, Covent Garden, London WC2H 9HE

Library of Congress Cataloging-in-Publication Data
Names: Datta, Supriyo, 1954–   author.
Title: Lessons from nanoelectronics : a new perspective on transport /
    Supriyo Datta, Purdue University, USA.
Other titles: Lessons from nanoscience ; v. 5.
Description: Second edition. | Singapore ; Hackensack, NJ : World Scientific, [2017]– |
    Series: Lessons from nanoscience: a lecture notes series, ISSN 2301-3354 ; vol. 5 |
    Includes bibliographical references and index.
    Contents: A. Basic concepts. Why electrons flow ; The elastic resistor ;
    Ballistic and diffusive transport ; Conductance from fluctuation ; Energy band model ;
    The nanotransistor ; Diffusion equation for ballistic transport ; Boltzmann equation ;
    Electrochemical potentials and quasi-Fermi levels ; Hall effect ; Smart contacts ;
    Thermoelectricity ; Phonon transport ; Second law ; Fuel value of information
Identifiers: LCCN 2016058007| ISBN 9789813224643 (set : alk. paper) |
    ISBN 9789813224650 (set : pbk. ; alk. paper) |
    ISBN 9789813209732 (Part A ; hardcover ; alk. paper) |
    ISBN 9813209739 (Part A ; hardcover ; alk. paper) |
    ISBN 9789813209749 (Part A; pbk. ; alk. paper) | ISBN 9813209747 (Part A ; pbk. ; alk. paper)
Subjects: LCSH: Nanoelectronics. | Transport theory.
Classification: LCC TK7874.84 .D37 2017 | DDC 621.381--dc23
LC record available at https://lccn.loc.gov/2016058007

Lessons from Nanoelectronics: A New Perspective on Transport (Volume 1)
ISBN: 9789814335287
ISBN: 9789814335294 (pbk)

British Library Cataloguing-in-Publication Data
A catalogue record for this book is available from the British Library.

Printed in Singapore

To Malika, Manoshi

and Anuradha

Preface

Everyone is familiar with the amazing performance of a modern smartphone, powered by a billion-plus nanotransistors, each having an active region that is barely a few hundred atoms long. I believe we also owe a major intellectual debt to the many who have made this technology possible.

This is because the same amazing technology has also led to a deeper understanding of the nature of current flow and heat dissipation on an atomic scale which I believe should be of broad relevance to the general problems of non-equilibrium statistical mechanics that pervade many different fields.

To make these lectures accessible to anyone in any branch of science or engineering, we assume very little background beyond linear algebra and differential equations. However, we will be discussing advanced concepts that should be of interest even to specialists, who are encouraged to look at my earlier books for additional technical details.

This book is based on a set of two online courses originally offered in 2012 on nanoHUB-U and more recently in 2015 on edX. In preparing the second edition we decided to split it into parts A and B entitled Basic Concepts and Quantum Transport respectively, along the lines of the two courses.

A list of available *video lectures* corresponding to different sections of this volume is provided upfront. I believe readers will find these useful.

Even this Second Edition represents lecture notes in unfinished form. I plan to keep posting additions/corrections at the companion website.

Acknowledgments

The precursor to this lecture note series, namely the *Electronics from the Bottom Up* initiative on www.nanohub.org was funded by the U.S. National Science Foundation (NSF), the Intel Foundation, and Purdue University. Thanks to World Scientific Publishing Corporation and, in particular, our series editor, Zvi Ruder for joining us in this partnership.

In 2012 nanoHUB-U offered its first two online courses based on this text. We gratefully acknowledge Purdue and NSF support for this program, along with the superb team of professionals who made nanoHUB-U a reality (https://nanohub.org/u) and later helped offer these courses through edX.

A special note of thanks to Mark Lundstrom for his leadership that made it all happen and for his encouragement and advice. I am grateful to Shuvro Chowdhury for carefully going through this new edition and fixing the many errors introduced during conversion to LaTeX. I also owe a lot to many students, ex-students, on-line students and colleagues for their valuable feedback and suggestions regarding these lecture notes.

Finally I would like to express my deep gratitude to all who have helped me learn, a list that includes many teachers, colleagues and students over the years, starting with the late Richard Feynman whose classic lectures on physics, I am sure, have inspired many like me and taught us the "pleasure of finding things out."

Supriyo Datta

List of Available Video Lectures

This book is based on a set of two online courses originally offered in 2012 on nanoHUB-U and more recently in 2015 on edX. These courses are now available in self-paced format at nanoHUB-U (https://nanohub.org/u) along with many other unique online courses.

Additional information about this book along with questions and answers is posted at the book website.

In preparing the second edition we decided to split the book into parts A and B following the two online courses available on nanoHUB-U entitled *Fundamentals of Nanoelectronics*

       Part A: Basic Concepts   Part B: Quantum Transport.

Video lecture of possible interest in this context: NEGF: A Different Perspective.

Following is a detailed list of *video lectures* available at the course website corresponding to different sections of this volume (Part A: Basic Concepts).

Constants Used in This Book

Electronic charge	$-q$	$=$	-1.6×10^{-19} C (Coulomb)
Unit of energy	1 eV	$=$	1.6×10^{-19} J (Joule)
Boltzmann constant	k	$=$	1.38×10^{-23} J$\cdot$K^{-1}
			~ 25 meV (at 300 K)
Planck's constant	h	$=$	6.626×10^{-34} J$\cdot$s
Reduced Planck's constant	$\hbar = h/2\pi$	$=$	1.055×10^{-34} J$\cdot$s
Free electron mass	m_0	$=$	9.109×10^{-31} kg

Some Symbols Used

m	Effective Mass	Kg
I	Electron Current	A (Amperes)
T	Temperature	K (Kelvin)
t	Transfer Time	s (second)
V	Electron Voltage	V (Volt)
U	Electrostatic Potential	eV
μ	Electrochemical Potential (also called Fermi level or quasi-Fermi level)	eV
μ_0	Equilibrium Electrochemical Potential	eV
R	Resistance	Ω (Ohm)
G	Conductance	S (Siemens)
$G(E)$	Conductance at 0 K with $\mu_0 = E$	S
λ	Mean Free Path for Backscattering	m
L_E	Energy Relaxation Length	m
L_{in}	Mean Path between Inelastic Scattering	m
τ_m	Momentum Relaxation time	s
$\overline{D}$	Diffusivity	$m^2 \cdot s^{-1}$
$\overline{\mu}$	Mobility	$m^2 \cdot V^{-1} \cdot s^{-1}$
ρ	Resistivity	$\Omega \cdot m$ (3D), Ω (2D), $\Omega \cdot m^{-1}$ (1D)
σ	Conductivity	$S \cdot m^{-1}$ (3D), S (2D), $S \cdot m$ (1D)
$\sigma(E)$	Conductivity at 0 K with $\mu_0 = E$	$S \cdot m^{-1}$ (3D), S (2D), $S \cdot m$ (1D)
A	Area	m^2

W	Width	m
L	Length	m
E	Energy	eV
C	Capacitance	F (Farad)
ϵ	Permittivity	$F \cdot m^{-1}$
ε	Energy	eV
$f(E)$	Fermi Function	Dimensionless
$\left(-\frac{\partial f}{\partial E}\right)$	Thermal Broadening Function (TBF)	eV^{-1}
$kT\left(-\frac{\partial f}{\partial E}\right)$	Normalized TBF	Dimensionless
$D(E)$	Density of States	eV^{-1}
$N(E)$	Number of States with Energy $< E$ (equals number of Electrons at 0 K with $\mu_0 = E$)	Dimensionless
n	Electron Density (3D or 2D or 1D)	m^{-3} or m^{-2} or m^{-1}
n_s	Electron Density	m^{-2}
n_L	Electron Density	m^{-1}
$M(E)$	Number of Channels (also called transverse modes)	Dimensionless
ν	Transfer Rate	s^{-1}
BTE	**B**oltzmann **T**ransport **E**quation	
NEGF	**N**on-**E**quilibrium **G**reen's **F**unction	
DOS	**D**ensity **O**f **S**tates	
QFL	**Q**uasi-**F**ermi **L**evel	

Contents

Simple Model for Density of States   57

Chapter 1

Overview

Related video lecture available at course website, Scientific Overview.

"Everyone" has a smartphone these days, and each smartphone has more than a billion transistors, making transistors more numerous than anything else we could think of. Even the proverbial ants, I am told, have been vastly outnumbered.

There are many types of transistors, but the most common one in use today is the Field Effect Transistor (FET), which is essentially a resistor consisting of a "channel" with two large contacts called the "source" and the "drain" (Fig. 1.1a).

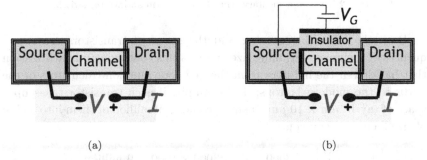

(a)   (b)

Fig. 1.1   (a) The Field Effect Transistor (FET) is essentially a resistor consisting of a channel with two large contacts called the source and the drain across which we attach the two terminals of a battery. (b) The resistance $R = V/I$ can be changed by several orders of magnitude through the gate voltage V_G.

The resistance (R) = Voltage (V)/Current (I) can be switched by several orders of magnitude through the voltage V_G applied to a third terminal

1

called the "gate" (Fig. 1.1b) typically from an "OFF" state of ~ 100 MΩ to an "ON" state of ~ 10 kΩ. Actually, the microelectronics industry uses a complementary pair of transistors such that when one changes from 100 MΩ to 10 kΩ, the other changes from 10 kΩ to 100 MΩ. Together they form an inverter whose output is the "inverse" of the input: a low input voltage creates a high output voltage while a high input voltage creates a low output voltage as shown in Fig. 1.2.

A billion such switches switching at GHz speeds (that is, once every nanosecond) enable a computer to perform all the amazing feats that we have come to take for granted. Twenty years ago computers were far less powerful, because there were "only" a million of them, switching at a slower rate as well.

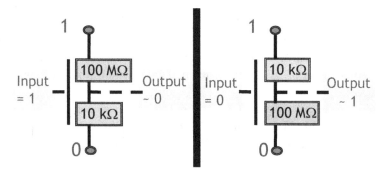

Fig. 1.2   A complementary pair of FET's form an inverter switch.

Both the increasing number and the speed of transistors are consequences of their ever-shrinking size and it is this continuing miniaturization that has driven the industry from the first four-function calculators of the 1970s to the modern laptops. For example, if each transistor takes up a space of say 10 μm $\times$ 10 μm, then we could fit 9 million of them into a chip of size 3 cm $\times$ 3 cm, since

$$\frac{3 \text{ cm}}{10 \ \mu\text{m}} = 3000 \quad \rightarrow \quad 3000 \times 3000 = 9 \text{ million.}$$

That is where things stood back in the ancient 1990s. But now that a transistor takes up an area of ~ 1 μm $\times$ 1 μm, we can fit 900 million (nearly a billion) of them into the same 3 cm $\times$ 3 cm chip. Where things will go from here remains unclear, since there are major roadblocks to continued miniaturization, the most obvious of which is the difficulty of dissipating the heat that is generated. Any laptop user knows how hot it gets when it is working hard, and it seems difficult to increase the number of switches or their speed too much further.

This book, however, is not about the amazing feats of microelectronics or where the field might be headed. It is about a less-appreciated by-product of the microelectronics revolution, namely the deeper understanding of current flow, energy exchange and device operation that it has enabled, which has inspired the perspective described in this book. Let me explain what we mean.

1.1 Conductance

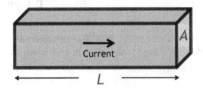

A basic property of a conductor is its resistance R which is related to the cross-sectional area A and the length L by the relation

$$R = \frac{V}{I} = \frac{\rho L}{A} \tag{1.1a}$$

$$G = \frac{I}{V} = \frac{\sigma A}{L}. \tag{1.1b}$$

The resistivity ρ is a geometry-independent property of the material that the channel is made of. The reciprocal of the resistance is the conductance G which is written in terms of the reciprocal of the resistivity called the conductivity σ. So what determines the conductivity?

Our usual understanding is based on the view of electronic motion through a solid as "diffusive" which means that the electron takes a random walk from the source to the drain, traveling in one direction for some length of time before getting scattered into some random direction as sketched in Fig. 1.3. The mean free path, that an electron travels before getting scattered is typically less than a micrometer (also called a micron = 10^{-3} mm, denoted μm) in common semiconductors, but it varies widely with temperature and from one material to another.

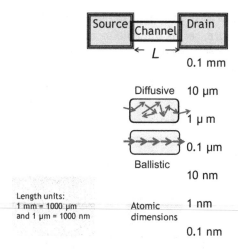

Fig. 1.3   The length of the channel of an FET has progressively shrunk with every new generation of devices ("Moore's law") and stands today at 14 nm, which amounts to ~ 100 atoms.

It seems reasonable to ask what would happen if a resistor is shorter than a mean free path so that an electron travels ballistically ("like a bullet") through the channel. Would the resistance still be proportional to length as described by Eq. (1.1a)? Would it even make sense to talk about its resistance?

These questions have intrigued scientists for a long time, but even twenty five years ago one could only speculate about the answers. Today the answers are quite clear and experimentally well established. Even the transistors in commercial laptops now have channel lengths $L \sim 14$ nm, corresponding to a few hundred atoms in length! And in research laboratories people have even measured the resistance of a hydrogen molecule.

1.2   Ballistic Conductance

It is now clearly established that the resistance R_B and the conductance G_B of a ballistic conductor can be written in the form

$$R_B = \frac{h}{q^2}\frac{1}{M} \;\simeq\; 25 \text{ k}\Omega \times \frac{1}{M} \tag{1.2a}$$

$$G_B = \frac{q^2}{h}M \simeq \; 40 \text{ }\mu\text{S} \times M \tag{1.2b}$$

where q, h are fundamental constants and M represents the number of effective channels available for conduction. Note that we are now using the word "channel" not to denote the physical channel in Fig. 1.3, but in the sense of parallel paths whose meaning will be clarified in the first two parts of this book. In future we will refer to M as the number of "modes", *a concept that is arguably one of the most important lessons of nanoelectronics and mesoscopic physics.*

1.3 What Determines the Resistance?

The ballistic conductance G_B (Eq. (1.2b)) is now fairly well-known, but the common belief is that it is relevant only for short conductors and belongs in a course on special topics like mesoscopic physics or nanoelectronics. We argue that the resistance for both long and short conductors can be written in terms of G_B (λ: mean free path)

$$G = \frac{G_B}{\left(1 + \dfrac{L}{\lambda}\right)}. \tag{1.3}$$

Ballistic and diffusive conductors are not two different worlds, but rather a continuum as the length L is increased. For $L \ll \lambda$, Eq. (1.3) reduces to $G \simeq G_B$, while for $L \gg \lambda$,

$$G \simeq \frac{G_B \lambda}{L},$$

which morphs into Ohm's law (Eq. (1.1b)) if we write the conductivity as

$$\sigma = \frac{GL}{A} = \frac{G_B}{A}\lambda = \frac{q^2}{h}\frac{M}{A}\lambda \quad (New\ Expression). \tag{1.4}$$

The conductivity of long diffusive conductors is determined by the number of modes per unit area (M/A) which represents a basic material property that is reflected in the conductance of ballistic conductors.

By contrast, the standard expressions for conductivity are all based on bulk material properties. For example freshman physics texts typically describe the Drude formula (momentum relaxation time: τ_m):

$$\sigma = q^2 \frac{n}{m} \tau_m \quad (Drude\ formula) \tag{1.5}$$

involving the effective mass (m) and the density of free electrons (n). This is the equation that many researchers carry in their head and use to interpret experimental data. However, it is tricky to apply if the electron dynamics is not described by a simple positive effective mass m. A more general but less well-known expression for the conductivity involves the density of states (D) and the diffusion coefficient $(\overline{D})$

$$\sigma = q^2 \frac{D}{AL} \overline{D} \quad (\textit{Degenerate Einstein relation}). \quad (1.6)$$

In Part 1 of this book we will use fairly elementary arguments to establish the new formula for conductivity given by Eq. (1.4) and show its equivalence to Eq. (1.6). In Part 2 we will introduce an energy band model and relate Eqs. (1.4) and (1.6) to the Drude formula (Eq. (1.5)) under the appropriate conditions when an effective mass can be defined.

We could combine Eqs. (1.3) and (1.4) to say that the standard Ohm's law (Eqs. (1.1)) should be replaced by the result

$$G = \frac{\sigma A}{L + \lambda} \rightarrow R = \frac{\rho}{A} (L + \lambda), \quad (1.7)$$

suggesting that the ballistic resistance (corresponding to $L \ll \lambda$) is equal to $\rho\lambda/A$ which is the resistance of a channel with resistivity ρ and length equal to the mean free path λ.

But this can be confusing since neither resistivity nor mean free path are meaningful for a ballistic channel. It is just that the resistivity of a diffusive channel is inversely proportional to the mean free path, and the product $\rho\lambda$ is a material property that determines the ballistic resistance R_B. A better way to write the resistance is from the inverse of Eq. (1.3):

$$R = R_B \left(1 + \frac{L}{\lambda} \right). \quad (1.8)$$

This brings us to a key conceptual question that caused much debate and discussion in the 1980s and still seems less than clear! Let me explain.

1.4   Where is the Resistance?

Equation (1.8) tells us that the total resistance has two parts

$$\underbrace{R_B}_{\text{length-independent}} \quad \text{and} \quad \underbrace{\frac{R_B L}{\lambda}}_{\text{length-dependent}} .$$

It seems reasonable to assume that the length-dependent part is associated with the channel. What is less clear is that the length-independent part (R_B) is associated with the interfaces between the channel and the two contacts as shown in Fig. 1.4.

How can we split up the overall resistance into different components and pinpoint them spatially? If we were talking about a large everyday resistor, the approach is straightforward: we simply look at the voltage drop across the structure. Since the same current flows everywhere, the voltage drop at any point should be proportional to the resistance at that point $\Delta V = I\Delta R$. A resistance localized at the interface should also give a voltage drop localized at the interface as shown in Fig. 1.4.

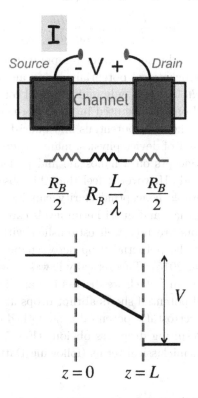

Fig. 1.4 The length-dependent part of the resistance in Eq. (1.8) is associated with the channel while the length-independent part is associated with the interfaces between the channel and the two contacts. Shown below is the spatial profile of the "potential" which supports the spatial distribution of resistances shown.

What makes this discussion not so straightforward in the context of nanoscale conductors is that it is not obvious how to draw a spatial potential profile on a nanometer scale. The key question is well-known in the context of electronic devices, namely the distinction between the electrostatic potential and the electrochemical potential.

The former is related to the electric field F

$$F = -\frac{d\phi}{dz},$$

since the force on an electron is qF, it seems natural to think that the current should be determined by $d\phi/dz$. However, it is well-recognized that this is only of limited validity at best. More generally current is driven by the gradient in the *electrochemical potential*:

$$\frac{I}{A} \equiv J = -\frac{\sigma}{q}\frac{d\mu}{dz}. \tag{1.9}$$

Just as heat flows from higher to lower temperatures, electrons flow from higher to lower electrochemical potentials giving an electron current that is proportional to $-d\mu/dz$. It is only under special conditions that μ and ϕ track each other and one can be used in place of the other. Although the importance of electrochemical potentials and quasi-Fermi levels is well established in the context of device physics, many experts feel uncomfortable about using these concepts on a nanoscale and prefer to use the electrostatic potential instead. However, I feel that this obscures the underlying physics and considerable conceptual clarity can be achieved by defining electrochemical potentials and quasi-Fermi levels carefully on a nanoscale.

The basic concepts are now well established with careful experimental measurements of the potential drop across nanoscale defects (see for example, Willke *et al.* 2015). Theoretically it was shown using a full quantum transport formalism (which we discuss in part B) that a suitably defined electrochemical potential shows abrupt drops at the interfaces, while the corresponding electrostatic potential is smoothed out over a screening length making the resulting drop less obvious (Fig. 1.5). These ideas are described in simple semiclassical terms (following Datta 1995) in Part 3 of this volume.

1.5   But Where is the Heat?

One often associates the electrochemical potential with the energy of the electrons, but at the nanoscale this viewpoint is completely incompatible with what we are discussing. The problem is easy to see if we consider an ideal ballistic channel with a defect or a barrier in the middle, which is the problem Rolf Landauer posed in 1957.

Common sense says that the resistance is caused largely by the barrier and we will show in Chapter 10 that a suitably defined electrochemical

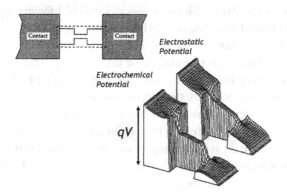

Fig. 1.5 Spatial profile of electrostatic and electrochemical potentials in a nanoscale conductor using a quantum transport formalism. Reproduced from *McLennan et al. 1991*.

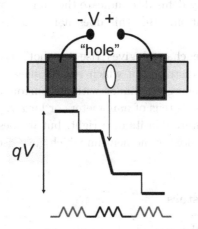

Fig. 1.6 Potential profile across a ballistic channel with a hole in the middle.

potential indeed shows a spatial profile that shows a sharp drop across the barrier in addition to abrupt drops at the interfaces as shown in Fig. 1.6.

If we associate this electrochemical potential with the energy of the electrons then an abrupt potential drop across the barrier would be accompanied by an abrupt drop in the energy, implying that heat is being dissipated locally at the scatterer. This requires the energy to be transferred from the electrons to the lattice so as to set the atoms jiggling which manifests itself as heat. But a scatterer does not necessarily have the degrees of freedom needed to dissipate energy: it could for example be just a hole in the middle of the channel with no atoms to "jiggle."

In short, the resistance R arises from the loss of momentum caused in this case by the "hole" in the middle of the channel. But the dissipation I^2R could occur very far from the hole and the potential in Fig. 1.6 cannot represent the energy. So what does it represent?

The answer is that the electrochemical potential represents the degree of filling of the available states, so that it indicates the number of electrons and not their energy. It is then easy to understand the abrupt drop across a barrier which represents a bottleneck on the electronic highway. As we all know there are traffic jams right before a bottleneck, but as soon as we cross it, the road is all empty: that is exactly what the potential profile in Fig. 1.6 indicates!

In short, everyone would agree that a "hole" in an otherwise ballistic channel is the cause and location of the resulting resistance and an electrochemical potential defined to indicate the number of electrons correlates well with this intuition. But this does not indicate the location of the dissipation I^2R.

The hole in the channel gives rise to "hot" electrons with a non-equilibrium energy distribution which relaxes back to normal through a complex process of energy exchange with the surroundings over an energy relaxation length $L_E \sim$ tens of nanometers or longer. The process of dissipation may be of interest in its own right, but it does not help locate the hole that caused the loss of momentum which gave rise to resistance in the first place.

1.6   Elastic Resistors

Once we recognize the spatially distributed nature of dissipative processes it seems natural to model nanoscale resistors shorter than L_E as an *ideal elastic resistor which we define as one in which all the energy exchange and dissipation occurs in the contacts and none within the channel itself* (Fig. 1.7).

For a ballistic resistor R_B, as my colleague Ashraf often points out, it is almost obvious that the corresponding Joule heat I^2R must occur in the contacts. After all a bullet dissipates most of its energy to the object it hits rather than to the medium it flies through.

There is experimental evidence that real nanoscale conductors do actually come close to this idealized model which has become widely used

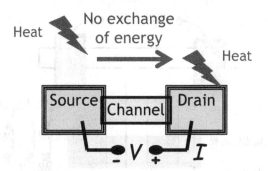

Fig. 1.7 *The ideal elastic resistor* with the Joule heat $VI = I^2R$ generated entirely in the contacts as sketched. Many nanoscale conductors are believed to be close to this ideal.

ever since the advent of mesoscopic physics in the late 1980s and is often referred to as the *Landauer approach*. However, it is generally believed that this viewpoint applies only to near-ballistic transport and to avoid this association we are calling it an *elastic resistor* rather than a *Landauer resistor*.

What we wish to stress is that even a diffusive conductor full of "potholes" that destroy momentum could in principle dissipate all the Joule heat in the contacts. And even if it does not, its resistance can be calculated accurately from an idealized model that assumes it does. Indeed we will use this elastic resistor model to obtain the conductivity expression in Eq. (1.4) and show that it agrees well with the standard results.

But surely we cannot ignore all the dissipation inside a long resistor and calculate its resistance accurately treating it as an elastic resistor? We believe we can do so in many cases of interest, especially at low bias. The underlying issues can be understood qualitatively using the simple circuit model shown in Fig. 1.8. For an elastic resistor each energy channel E_1, E_2 and E_3 is independent with no flow of electrons between them as shown on the left. Inelastic processes induce "vertical" flow between the energy channels represented by the vertical resistors as shown on the right. When can we ignore the vertical resistors?

If the series of resistors representing individual channels are identical, then the nodes connected by the vertical resistors will be at the same potential, so that *there will be no current flow through them*. Under these conditions, an elastic resistor model that ignores the vertical resistors is quite accurate.

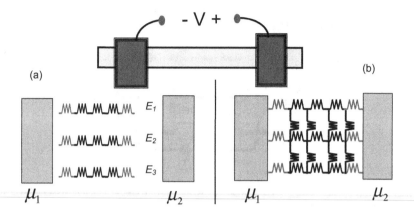

Fig. 1.8   A simple circuit model: (a) For elastic resistors, individual energy channels E_1, E_2 and E_3 are decoupled with no flow between them. (b) Inelastic processes cause vertical flow between energy channels through the additional resistors shown.

But vertical flow cannot always be ignored. For example, Fig. 1.9a shows a conductor where the lower energy levels E_2 and E_3 conduct poorly compared to E_1. We would then expect the electrons to flow upwards in energy on the left and downwards in energy on the right as shown, thus cooling the lattice on the left and heating the lattice on the right, leading to the well-known *Peltier effect* discussed in Chapter 13.

The role of vertical flow can be even more striking if the left contact connects only to the channel E_1 while the right contact connects only to E_3 as shown in Fig. 1.9b. No current can flow in such a structure without vertical flow, and the entire current is purely a vertical current. This is roughly what happens in p-n junctions which is discussed a little further in Section 12.1.

The bottom line is that elastic resistors generally provide a good description of short conductors and the Landauer approach has become quite common in mesoscopic physics and nanoelectronics. What is not well recognized is that this approach can provide useful results even for long conductors. In many cases, but not always, we can ignore inelastic processes and calculate the resistance quite accurately as long as the momentum relaxation has been correctly accounted for, as discussed further in Section 3.3.

But why would we want to ignore inelastic processes? Why is the theory of elastic resistors any more straightforward than the standard approach? To understand this we first need to talk briefly about the transport theories

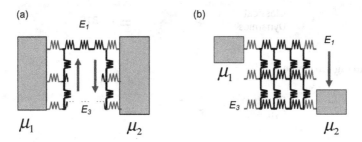

Fig. 1.9 Two examples of structures where vertical flow between energy channels can be important: (a) If the lower energy levels E_2 and E_3 conduct poorly, electrons will flow up in energy on the left and down in energy on the right as shown. (b) If the left contact couples to an upper energy E_1 while the right contact couples to a lower energy E_3, then the current flow is purely vertical, occurring *only* through inelastic processes.

on which the standard approach is based.

1.7 Transport Theories

Flow or transport always involves two fundamentally different types of processes, namely elastic transfer and heat generation, belonging to two distinct branches of physics. The first involves frictionless mechanics of the type described by Newton's laws or the Schrödinger equation. The second involves the generation of heat described by the laws of thermodynamics.

The first is driven by forces or potentials and is reversible. The second is driven by entropy and is irreversible. Viewed in reverse, entropy-driven processes look absurd, like heat flowing spontaneously from a cold to a hot surface or an electron accelerating spontaneously by absorbing heat from its surroundings.

Normally the two processes are intertwined and a proper description of current flow in electronic devices requires the advanced methods of non-equilibrium statistical mechanics that integrate mechanics with thermodynamics. Over a century ago Boltzmann taught us how to combine Newtonian mechanics with heat generating or entropy-driven processes and the resulting Boltzmann transport equation (BTE) is widely accepted as the cornerstone of semiclassical transport theory. The word semiclassical is used because some quantum effects have also been incorporated approximately into the same framework.

A full treatment of quantum transport requires a formal integration of quantum dynamics described by the Schrödinger equation with heat

Classical
Dynamics   **+** ⚡ **=** BTE

generating processes.

Quantum
Dynamics   **+** ⚡ **=** NEGF

This is exactly what is achieved in the non-equilibrium Green's function (NEGF) method originating in the 1960s from the seminal works of Martin and Schwinger (1959), Kadanoff and Baym (1962), Keldysh (1965) and others.

1.7.1 *Why elastic resistors are conceptually simpler*

The BTE takes many semesters to master and the full NEGF formalism, even longer. Much of this complexity comes from the subtleties of combining mechanics with distributed heat-generating processes.

The operation of the elastic resistor can be understood in far more elementary terms because of the clean spatial separation between the force-driven and the entropy-driven processes. The former is confined to the channel and the latter to the contacts. As we will see in the next few chapters, the latter is easily taken care of, indeed so easily that it is easy to miss the profound nature of what is being accomplished.

Even quantum transport can be discussed in relatively elementary terms using this viewpoint. For example, Fig. 1.10 shows a plot of the spatial profile of the electrochemical potential across our structure from Fig. 1.6 with a hole in the middle, calculated both from the semiclassical BTE (Chapter 9) and from the NEGF method (part B).

For the NEGF method we show three options. First a coherent model (left) that ignores all interaction within the channel showing oscillations indicative of standing waves. Once we include phase relaxation, the constructive and destructive interferences are lost and we obtain the result in

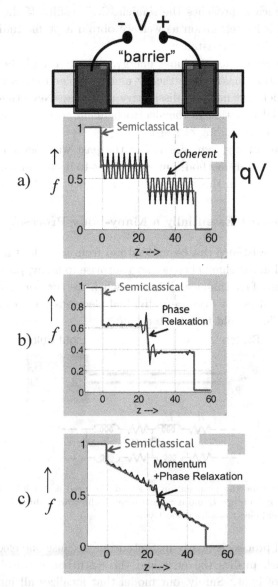

Fig. 1.10 Spatial profile of the electrochemical potential across a channel with a barrier. Solid red line indicates semiclassical result from BTE (part A). Also shown are the results from NEGF (part B) assuming (a) coherent transport, (b) transport with phase relaxation, (c) transport with phase and momentum relaxation. Note that no energy relaxation is included in any of these calculations.

the middle which approaches the semiclassical result. If the interactions include momentum relaxation as well we obtain a profile indicative of an additional distributed resistance.

None of these models includes energy relaxation and they all qualify as elastic resistors making the theory much simpler than a full quantum transport model that includes dissipative processes. Nevertheless, they all exhibit a spatial variation in the electrochemical potential consistent with our intuitive understanding of resistance.

A good part of my own research in the past was focused in this area developing the NEGF method, but we will get to it only in part B after we have "set the stage" in this volume using a semiclassical picture.

1.8 Is Transport Essentially a Many-body Process?

The idea that resistance can be understood from a model that ignores interactions within the channel comes as a surprise to many, possibly because of an interesting fact that we all know: when we turn on a switch and a bulb lights up, it is not because individual electrons flow from the switch to the bulb. That would take far too long.

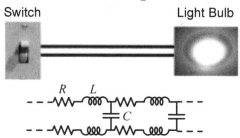

Fig. 1.11 To describe the propagation of signals we need a distributed RLC, model that includes an inductance L and a capacitance C which are ordinarily determined by magnetostatics and electrostatics respectively.

The actual process is nearly instantaneous because one electron pushes the next, which pushes the next and the disturbance travels essentially at the speed of light. Surely, our model that localizes all interactions at arbitrarily placed contacts (Fig. 3.5) cannot describe this process?

The answer is that to describe the propagation of transient signals we need a model that includes not just a resistance R, but also an inductance L and a capacitance C as shown in Fig. 1.11. These could include transport related corrections in small conductors but are ordinarily determined by magnetostatics and electrostatics respectively (Salahuddin *et al.* 2005).

In this distributed RLC transmission line, the signal velocity determined by L and C can be well in excess of individual electron velocities reflecting a collective process. However, L and C play no role at low frequencies, since the inductor is then like a "short circuit" and the capacitor is like an "open circuit." The low frequency conduction properties are represented solely by the resistance R and can usually be understood fairly well in terms of the transport of individual electrons along M parallel modes (see Eqs. (1.2)) or "channels", a concept that has emerged from decades of research. To quote Phil Anderson from a volume commemorating 50 years of Anderson localization (see Anderson (2010)):

" ... *What might be of modern interest is the "channel" concept which is so important in localization theory. The transport properties at low frequencies can be reduced to a sum over one-dimensional "channels" ...* "

1.9 A Different Physical Picture

Let me conclude this overview with an obvious question: why should we bother with idealized models and approximate physical pictures? Can't we simply use the BTE and the NEGF equations which provide rigorous frameworks for describing semiclassical and quantum transport respectively? The answer is yes, and all the results we discuss are benchmarked against the BTE and the NEGF.

However, as Feynman (1963) noted in his classic lectures, even when we have an exact mathematical formulation, we need an intuitive physical picture:

"*.. people .. say .. there is nothing which is not contained in the equations .. if I understand them mathematically inside out, I will understand the physics inside out. Only it doesn't work that way. .. A physical understanding is a completely unmathematical, imprecise and inexact thing, but absolutely necessary for a physicist.*"

Indeed, most researchers carry a physical picture in their head and it is usually based on the Drude formula (Eq. (1.5)). In this book we will show that an alternative picture based on elastic resistors leads to a formula (Eq. (1.4)) that is more generally valid.

Unlike the Drude formula which treats the electric field as the driving term, this new approach more correctly treats the electrochemical potential as the driving term. This is well-known at the macroscopic level, but somehow seems to have been lost in nanoscale transport, where people cite the difficulty of defining electrochemical potentials. However, that does not justify using electric field as a driving term, an approach that *does not work for inhomogeneous conductors on any scale*.

Since all conductors are fundamentally inhomogeneous on an atomic scale it seems questionable to use electric field as a driving term. We argue that at least for low bias transport, it is possible to define electrochemical potentials or quasi-Fermi levels on an atomic scale and this can lend useful insight into the physics of current flow and the origin of resistance. We believe this is particularly timely because future electronic devices will require a clear understanding of the different potentials.

For example, recent work on spintronics has clearly established experimental situations where upspin and downspin electrons have different electrochemical potentials (sometimes called quasi-Fermi levels) and could even flow in opposite directions because their $d\mu/dz$ have opposite signs. This cannot be understood if we believe that currents are driven by electric fields, $-d\phi/dz$, since up and down spins both see the same electric field and have the same charge. We can expect to see more and more such examples that use novel contacts to manipulate the quasi-Fermi levels of different group of electrons (see Chapter 12 for further discussion).

In short we believe that the lessons of nanoelectronics lead naturally to a new viewpoint, one that changes even some basic concepts we all learn in freshman physics. This viewpoint represents a departure from the established mindset and I hope it will provide a complementary perspective to facilitate the insights needed to take us to the next level of discovery and innovation.

PART 1
What Determines the Resistance

Chapter 2

Why Electrons Flow

It is a well-known and well-established fact that when the two terminals of a battery are connected across a conductor, it gives rise to a current due to the flow of electrons across the channel from the source to the drain.

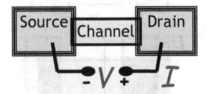

If you ask anyone, novice or expert, what causes electrons to flow, by far the most common answer you will receive is that it is the electric field. However, this answer is incomplete at best. After all even before we connect a battery, there are enormous electric fields around every atom due to the positive nucleus whose effects on the atomic spectra are well-documented. Why is it that these electric fields do not cause electrons to flow, and yet a far smaller field from an external battery does?

The standard answer is that microscopic fields do not cause current to flow, a macroscopic field is needed. This too is not satisfactory for two reasons. Firstly, there are well-known inhomogeneous conductors like p-n junctions which have large macroscopic fields extending over many micrometers that do not cause any flow of electrons till an external battery is connected.

Secondly, experimentalists are now measuring current flow through conductors that are only a few atoms long with no clear distinction between the microscopic and the macroscopic. This is a result of our progress in nanoelectronics, and it forces us to search for a better answer to the question, "why electrons flow."

2.1   Two Key Concepts

Related video lecture available at course website, Unit 1: L1.2.

To answer this question, we need two key concepts. First is the *density of states per unit energy $D(E)$ available for electrons to occupy* inside the channel (Fig. 2.1). For the benefit of experts, I should note that we are adopting what we will call a "point channel model" represented by a single density of states $D(E)$. More generally one needs to consider the spatial variation of $D(E)$, as we will see in Chapter 7, but there is much that can be understood just from our point channel model.

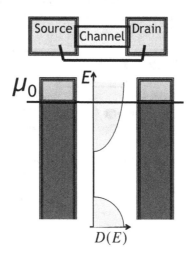

Fig. 2.1   The first step in understanding the operation of any electronic device is to draw the available density of states $D(E)$ as a function of energy E, inside the channel and to locate the equilibrium electrochemical potential μ_0 separating the filled from the empty states.

The second key input is the *location of the electrochemical potential, μ_0* which at equilibrium is the same everywhere, in the source, in the drain, and in the channel. Roughly speaking (we will make this statement more precise shortly) it is the energy that demarcates the filled states from the empty ones. All states with energy $E < \mu_0$ are filled while all states with $E > \mu_0$ are empty. For convenience I might occasionally refer to the electrochemical potential as just the "potential".

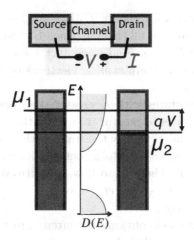

Fig. 2.2   When a voltage is applied across the contacts, it lowers all energy levels at the positive contact (drain in the picture). As a result the electrochemical potentials in the two contacts separate: $\mu_1 - \mu_2 = qV$.

When a battery is connected across the two contacts creating a potential difference V between them, it lowers all energies at the positive terminal (drain) by an amount qV, $-q$ being the charge of an electron ($q = 1.6 \times 10^{-19}$ C) separating the two electrochemical potentials by qV as shown in Fig. 2.2:

$$\mu_1 - \mu_2 = qV. \tag{2.1}$$

Just as a temperature difference causes heat to flow and a difference in water levels makes water flow, a difference in electrochemical potentials causes electrons to flow. Interestingly, only the states in and around an energy window around μ_1 and μ_2 contribute to the current flow, all the states far above and well below that window playing no part at all. Let me explain why.

2.1.1   *Energy window for current flow*

Each contact seeks to bring the channel into equilibrium with itself, which roughly means filling up all the states with energies E less than its electrochemical potential μ and emptying all states with energies greater than μ.

Consider the states with energy E that are less than μ_1 but greater than μ_2. Contact 1 wants to fill them up since $E < \mu_1$, but contact 2 wants to empty them since $E > \mu_2$. And so contact 1 keeps filling them up and contact 2 keeps emptying them causing electrons to flow continually from contact 1 to contact 2.

Consider now the states with E greater than both μ_1 and μ_2. Both contacts want these states to remain empty and they simply remain empty with no flow of electrons. Similarly the states with E less than both μ_1 and μ_2 do not cause any flow either. Both contacts like to keep them filled and they just remain filled. There is no flow of electrons outside the window between μ_1 and μ_2, or more correctly outside $\pm$ a few kT of this window, as we will discuss shortly.

This last point may seem obvious, but often causes much debate because of the common belief we alluded to earlier, namely that electron flow is caused by the electric field in the channel. If that were true, all the electrons should flow and not just the ones in any specific window determined by the contacts.

2.2   Fermi Function

Let us now make the above statements more precise. We stated that roughly speaking, at equilibrium, all states with energies E below the electrochemical potential μ are filled while all states with $E > \mu$ are empty. This is precisely true only at absolute zero temperature. More generally, the transition from completely full to completely empty occurs over an energy range $\sim \pm 2\,kT$ around $E = \mu$ where k is the Boltzmann constant (~ 80 μeV/K) and T is the absolute temperature. Mathematically, this transition is described by the Fermi function:

$$f(E) = \frac{1}{\exp\left(\dfrac{E - \mu}{kT}\right) + 1}. \tag{2.2}$$

This function is plotted in Fig. 2.3 (left panel), though in an unconventional form with the energy axis vertical rather than horizontal. This will allow us to place it alongside the density of states, when trying to understand current flow (see Fig. 2.4).

For readers unfamiliar with the Fermi function, let me note that an extended discussion is needed to do justice to this deep but standard result, and we will discuss it a little further in Chapter 15 when we talk about the key principles of equilibrium statistical mechanics. At this stage it may

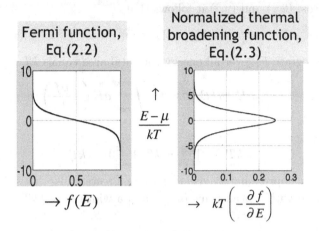

Fig. 2.3 Fermi function and the normalized (dimensionless) thermal broadening function.

help to note that what this function (Fig. 2.3) basically tells us is that states with low energies are always occupied ($f = 1$), while states with high energies are always empty ($f = 0$), something that seems reasonable since we have heard often enough that (1) everything goes to its lowest energy, and (2) electrons obey an exclusion principle that stops them from all getting into the same state. The additional fact that the Fermi function tells us is that the transition from $f = 1$ to $f = 0$ occurs over an energy range of $\sim \pm 2\,kT$ around μ_0.

2.2.1 *Thermal broadening function*

Also shown in Fig. 2.3 is the derivative of the Fermi function, multiplied by kT to make it dimensionless. Using Eq. (2.2) it is straightforward to show that

$$F_T(E, \mu) = kT\left(-\frac{\partial f}{\partial E}\right) = \frac{e^x}{(e^x + 1)^2} \tag{2.3}$$

where $\quad x \equiv \dfrac{E - \mu}{kT}.$

Note: (1) From Eq. (2.3) it follows that

$$F_T(E, \mu) = F_T(E - \mu) = F_T(\mu - E). \tag{2.4}$$

(2) From Eqs. (2.3) and (2.2) it follows that

$$F_T = f(1 - f).$$ (2.5)

(3) If we integrate F_T over all energy the total area equals kT:

$$\int_{-\infty}^{+\infty} dE\ F_T(E, \mu) = kT \int_{-\infty}^{+\infty} dE \left(-\frac{\partial f}{\partial E} \right)$$

$$= kT\ [-f]_{-\infty}^{+\infty} = kT(1 - 0) = kT$$ (2.6)

so that we can approximately visualize F_T as a rectangular "pulse" centered around $E = \mu$ with a peak value of $1/4$ and a width of $\sim 4\,kT$.

2.3 Non-equilibrium: Two Fermi Functions

When a system is in equilibrium the electrons are distributed among the available states according to the Fermi function. But when a system is driven out-of-equilibrium there is no simple rule for determining the distribution of electrons. It depends on the specific problem at hand making non-equilibrium statistical mechanics far richer and less understood than its equilibrium counterpart.

For our specific non-equilibrium problem, we argue that the two contacts are such large systems that they cannot be driven out-of-equilibrium. And so each remains locally in equilibrium with its own electrochemical potential giving rise to two different Fermi functions (Fig. 2.4):

$$f_1(E) = \frac{1}{\exp\left(\dfrac{E - \mu_1}{kT}\right) + 1}$$ (2.7a)

$$f_2(E) = \frac{1}{\exp\left(\dfrac{E - \mu_2}{kT}\right) + 1}.$$ (2.7b)

The "little" channel in between does not quite know which Fermi function to follow and as we discussed earlier, the source keeps filling it up while the drain keeps emptying it, resulting in a continuous flow of current.

In summary, what makes electrons flow is the difference in the "agenda" of the two contacts as reflected in their respective Fermi functions, $f_1(E)$ and $f_2(E)$. This is qualitatively true for all conductors, short or long. But for short conductors, the current at any given energy E is quantitatively proportional to

$$I(E) \sim f_1(E) - f_2(E) \tag{2.8}$$

representing the difference in the occupation probabilities in the two contacts. This quantity goes to zero when E lies way above μ_1 and μ_2, since f_1 and f_2 are both zero. It also goes to zero when E lies way below μ_1 and μ_2, since f_1 and f_2 are both one. Current flow occurs only in the intermediate energy window, as we had argued earlier.

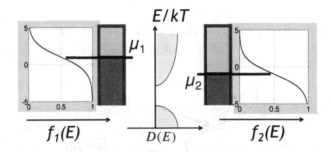

Fig. 2.4   Electrons in the contacts occupy the available states with a probability described by a Fermi function $f(E)$ with the appropriate electrochemical potential μ.

2.4   Linear Response

Current-voltage relations are typically not linear, but there is a common approximation that we will frequently use throughout this book to extract the "linear response" which refers to the low bias conductance, dI/dV, as $V \to 0$. The basic idea can be appreciated by plotting the difference between two Fermi functions, normalized to the applied voltage

$$F(E) = \frac{f_1(E) - f_2(E)}{qV/kT} \tag{2.9}$$

where

$$\mu_1 = \mu_0 + (qV/2)$$

$$\mu_2 = \mu_0 - (qV/2).$$

Figure 2.5 shows that the difference function F gets narrower as the voltage is reduced relative to kT. The interesting point is that as qV is reduced below kT, the function F approaches the thermal broadening function F_T we defined (see Eq. (2.3)) in Section 2.2.1:

$$F(E) \rightarrow F_T(E), \quad \text{as} \quad qV/kT \rightarrow 0$$

so that from Eq. (2.9)

$$f_1(E) - f_2(E) \approx \frac{qV}{kT} F_T(E, \mu_0) = \left(-\frac{\partial f_0}{\partial E} \right) qV \tag{2.10}$$

if the applied voltage $\mu_1 - \mu_2 = qV$ is much less than kT.

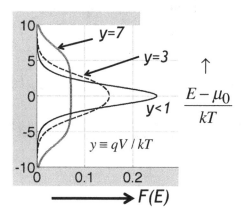

Fig. 2.5 $F(E)$ from Eq. (2.9) versus $(E - \mu_0)/kT$ for different values of $y = qV/kT$.

The validity of Eq. (2.10) for $qV \ll kT$ can be checked numerically if you have access to MATLAB or equivalent. For those who like to see a mathematical derivation, Eq. (2.10) can be obtained using the Taylor series expansion described in Appendix A to write

$$f(E) - f_0(E) \approx \left(-\frac{\partial f_0}{\partial E} \right) (\mu - \mu_0) \tag{2.11}$$

Eq. (2.11) and Eq. (2.10) which follows from it, will be used frequently in these lectures.

2.5 Difference in "Agenda" Drives the Flow

Before moving on, let me quickly reiterate the key point we are trying to make, namely that the current is determined by

$$-\frac{\partial f_0(E)}{\partial E} \quad \text{and } NOT \text{ by } \quad f_0(E).$$

The two functions look similar over a limited range of energies

$$-\frac{\partial f_0(E)}{\partial E} \approx \frac{f_0(E)}{kT} \quad \text{if } \ E - \mu_0 \gg kT.$$

So if we are dealing with a so-called non-degenerate conductor (see Section 3.4) where we can restrict our attention to a range of energies satisfying this criterion, we may not notice the difference.

In general these functions look very different (see Fig. 2.3) and the experts agree that current depends not on the Fermi function, but on its derivative. However, we are not aware of an elementary treatment that leads to this result and consequently our everyday thinking tends to be dominated by a different picture.

2.5.1 *Drude formula*

Related video lecture available at course website, Unit 1: L1.9.

For example, freshman physics texts start by treating the force due to an electric field F as the driving term and adding a frictional term to Newton's law (τ_m is the so-called "momentum relaxation time")

$$\underbrace{\frac{d(mv)}{dt} = (-qF)}_{Newton's\ Law} \quad \underbrace{- \quad \frac{mv}{\tau_m}}_{Friction}.$$

At steady-state ($d/dt = 0$) this gives a non-zero drift velocity,

$$v_d = - \underbrace{\frac{q\,\tau_m}{m}}_{\text{mobility, } \tilde{\mu}} \times F \qquad (2.12)$$

from which one calculates the electron current using the relation

$$\frac{I}{A} = qnv_d = - \underbrace{qn\tilde{\mu}}_{\text{conductivity, } \sigma} \times F. \qquad (2.13)$$

The negative sign appears in Eq. (2.13) because we use "I" to denote the electron current (see Section 3.2.2) which flows opposite to the electric field.

Equations (2.12) and (2.13) lead to the Drude formula, stated earlier in Eq. (1.5), which plays a key role in defining our mental picture of current flow. Since the above approach treats electric fields as the driving term, it also suggests that the current depends on the total number of electrons since all electrons feel the field. This is commonly explained away by saying that there are mysterious quantum mechanical forces that prevent electrons in full bands from moving and what matters is the number of "free electrons". But this begs the question of which electrons are free and which are not, a question that becomes more confusing for atomic scale conductors.

It is well-known that the conductivity varies widely, changing by a factor of $\sim 10^{20}$ going from copper to glass, to mention two materials that are near two ends of the spectrum. But this is not because one has more electrons than the other. The total number of electrons is of the same order of magnitude for all materials from copper to glass. Whether a material is a good or a bad conductor is determined by the availability of states in an energy window $\sim kT$ around the electrochemical potential μ_0, which can vary widely from one material to another. This is well-known to experts and comes mathematically from the dependence of the conductivity

$$\text{on} \quad -\frac{\partial f_0}{\partial E} \quad \text{rather than } f_0(E)$$

a result that typically requires advanced treatments based on the Boltzmann equation (Chapter 9) or the fluctuation-dissipation theorem (Chapter 5).

2.5.2 *Present approach*

We obtain this result in an elementary way as we have just seen. Current is driven by the difference in the "agenda" of the two contacts which for low bias is proportional to the derivative of the equilibrium Fermi function:

$$f_1(E) - f_2(E) \approx \left(-\frac{\partial f_0}{\partial E}\right) qV.$$

There is no need to invoke mysterious forces that stops some electrons from moving, though one could perhaps call it a mysterious force, since the Fermi function (Eq. (2.2)) reflects the exclusion principle.

Later when we (briefly) discuss phonon transport in Chapter 14, we will see how this approach is readily extended to describe the flow of phonons. The phonon current is governed by the Bose (not Fermi) function which is appropriate for particles that do not have an exclusion principle.

Chapter 3

The Elastic Resistor

Related video lectures available at course website, Unit 1: L1.3 and L1.4.

We saw in the last chapter that the flow of electrons is driven by the difference in the "agenda" of the two contacts as reflected in their respective Fermi functions, $f_1(E)$ and $f_2(E)$. The negative contact with its larger $f(E)$ would like to see more electrons in the channel than the positive contact. And so the positive contact keeps withdrawing electrons from the channel while the negative contact keeps pushing them in. This is true of all conductors, big and small. But in general, it is difficult to express the current as a simple function of $f_1(E)$ and $f_2(E)$, because electrons jump around from one energy to another and the current flow at different energies is all mixed up.

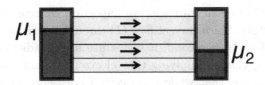

Fig. 3.1   An elastic resistor: electrons travel along fixed energy channels.

But for the ideal elastic resistor shown in Fig. 3.1, the current in an energy range from E to $E + dE$ is decoupled from that in any other energy range, allowing us to write it in the form

$$dI \sim dE \, G(E) \, (f_1(E) - f_2(E))$$

and integrating it to obtain the total current I. Making use of Eq. (2.10), this leads to an expression for the low bias conductance

$$\frac{I}{V} = \int_{-\infty}^{+\infty} dE \left(-\frac{\partial f_0}{\partial E} \right) G(E) \tag{3.1}$$

where $-(\partial f_0/\partial E)$ can be visualized as a rectangular pulse of area equal to one, with a width of $\sim 4\,kT$ around μ (see Fig. 2.3, right panel).

Equation (3.1) tells us that for an elastic resistor, we can define a conductance function $G(E)$ whose average over an energy range $\sim \pm 2kT$ around the electrochemical potential μ_0 gives the experimentally measured conductance. At low temperatures, we can simply use the value of $G(E)$ at $E = \mu_0$.

This energy-resolved view of conductance represents an enormous simplification that is made possible by the concept of an elastic resistor. Note that by elastic we do not just mean ballistic which implies that the electron goes "like a bullet" from source to drain. An electron could also take a more traditional diffusive path *as long as it changes only its momentum and not its energy along the way:*

(a)   (b)

In Section 3.1, we will start with an important conceptual issue regarding elastic resistors: Since current flow (I) through a resistor (R) dissipates a Joule heat of I^2R per second, it seems like a contradiction to talk of a resistor as being elastic, implying that electrons do not lose any energy. The point to note is that while the electron does not lose any energy in the channel of an elastic resistor, it does lose energy both in the source and the drain and that is where the Joule heat gets dissipated. *In short an elastic resistor has a resistance R determined by the channel, but the corresponding heat I^2R is entirely dissipated outside the channel.* This is a very non-intuitive result that seems to be at least approximately true of nanoscale conductors.

We will argue that it also helps understand transport properties like the conductivity of large resistors by viewing them as multiple elastic resistors in series making it a very powerful conceptual tool for transport problems in general. In Section 3.2 we will proceed to obtain our conductance formula

$$G(E) = \frac{q^2 D(E)}{2t(E)} \tag{3.2}$$

expressing the conductance function $G(E)$ for an elastic resistor in terms of the density of states $D(E)$ and the time $t(E)$ that an electron spends in the channel. Equation (3.2) seems quite intuitive: it says that the conductance is proportional to the product of two factors, namely *the availability of states (D) and the ease with which electrons can transport through them* $(1/t)$. This is the key result that we will use in subsequent chapters.

Finally in Section 3.3 we will explain how the energy averaging of the conductance function $G(E)$ described by Eq. (3.1) leads to two different limiting physical pictures, generally referred to as the degenerate and the non-degenerate limits. Although the semiconductor literature often focuses on the non-degenerate limit, we will try to keep the discussion general so that it applies to both limits and in-between as well.

3.1 How an Elastic Resistor Dissipates Heat

Let us start by addressing a basic question regarding an elastic resistor. How does it dissipate the joule heat I^2R associated with the resistance R?

Consider a one level elastic resistor having one sharp level with energy ε. Every time an electron crosses over through the channel, it appears as a "hot electron" on the drain side with an energy ε in excess of the local electrochemical potential μ_2 as shown below:

(a) Temporary state immediately after electron transfer

(b) Final state after energy relaxation processes have returned contacts to equilibrium

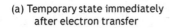

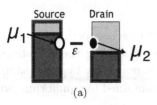

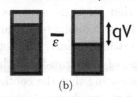

(a)

(b)

Energy dissipating processes in the contact quickly make the electron get rid of the excess energy $(\varepsilon - \mu_2)$. Similarly at the source end an empty spot (a "hole") is left behind with an energy that is much less than the local electrochemical potential μ_1, which gets quickly filled up by electrons dissipating the excess energy $(\mu_1 - \varepsilon)$.

In effect, every time an electron crosses over from the source to the drain,

an energy $(\mu_1 - \varepsilon)$ *is dissipated at the source*

an energy $(\varepsilon - \mu_2)$ *is dissipated at the drain.*

The total energy dissipated is

$$\mu_1 - \mu_2 = qV$$

which is supplied by the external battery that maintains the potential difference $\mu_1 - \mu_2$. The overall flow of electrons and heat is summarized in Fig. 3.2.

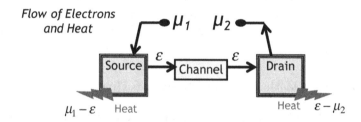

Fig. 3.2   Flow of electrons and heat in a one-level elastic resistor having one level with $E = \varepsilon$.

If N electrons cross over in a time t

$$\text{Current, } I = \frac{qN}{t}$$

$$\text{Dissipated power} = \frac{qVN}{t} = VI.$$

Note that VI is the same as I^2R and V^2G.

The heat dissipated by an "elastic resistor" thus occurs in the contacts. As we will see next, the detailed mechanism underlying the complicated process of heat transfer in the contacts can be completely bypassed simply by legislating that the contacts are always maintained in equilibrium with a fixed electrochemical potential.

3.2   Current in an Elastic Resistor

Consider an elastic resistor (Fig. 3.1) with an arbitrary density of states $D(E)$. Since all energy channels conduct independently in parallel, we could first write the current in an energy channel between E and $E + dE$ and then integrate over energy to find the total current.

To write the current in this energy range let us first assume that $f_1 = 1$ and $f_2 = 0$, so that electrons continually flow from 1 to 2. The flux of electrons per second can be related to the steady-state number of electrons in the channel as follows:

$$Electron\ flux\ =\ \frac{Number\ of\ electrons\ in\ channel}{Time\ each\ electron\ spends\ in\ channel}.$$

This is a non-obvious result, but one that appears in many different physical problems. For example we could relate the "flux" of students graduating each year from a given program to the number of students in the program by the relation

$$Student\ flux\ =\ \frac{Number\ of\ students\ in\ program}{Time\ each\ student\ spends\ in\ program}.$$

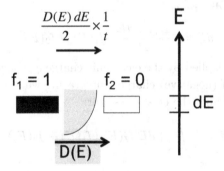

Using this principle we could write the flux of electrons as

$$\frac{D(E)\ dE}{2}\ \frac{1}{t}$$

where t is the *time an average electron spends in the channel* on its way from contact 1 to contact 2. This is because the steady-state number of electrons in the channel is equal to half the number of available states $D(E)\ dE$ since one contact wants to keep all states filled ($f_1 = 1$) while the other wants to keep it empty ($f_2 = 0$). On the average the states remain filled only half the time.

Interestingly, this half-filling of available states applies to both ballistic and diffusive regimes and in between, though the details are somewhat

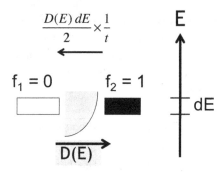

different in different regimes as we will see later in Chapter 8 when we discuss the diffusion equation.

If we reverse the roles of the two contacts with $f_1 = 0$ and $f_2 = 1$, we will have the same flux but in the opposite direction. In general with arbitrary values of f_1 and f_2 we can superpose the two results to write the net current as the difference

$$dI = q \, \frac{D(E) \, dE}{2t} \, (f_1(E) - f_2(E))$$

where we have multiplied by the electronic charge q to convert from flux of electrons to flux of (negative) charge. Integrating we obtain an expression for the current through the elastic resistor:

$$I = \frac{1}{q} \int_{-\infty}^{+\infty} dE \; G(E) \; (f_1(E) - f_2(E)) \tag{3.3}$$

where $G(E) = \dfrac{q^2 D(E)}{2t(E)} \quad \rightarrow \quad$ same as Eq. (3.2).

3.2.1 *Exclusion principle?*

It is not obvious that in general we can superpose the two results from $f_1 = 1, f_2 = 0$ and from $f_1 = 0, f_2 = 1$ and obtain the resulting current as we have done. Doesn't the presence of one stream affect the other?

We view electrons as non-interacting particles with the understanding that their Coulomb interaction is included through an average potential U which is calculated self-consistently (Chapter 7). But even then it might seem that the mere presence of one stream could impede the other stream through the Pauli exclusion principle which can be significant for degenerate conductors (see Section 3.4) where f is close to one in the energy range of interest.

However, it can be shown that the two flows involve "orthogonal states" that do not "Pauli block" each other in any way, as long as transport is coherent and does not involve interaction with external objects having internal degrees of freedom (see Section 2.6 of Datta (1995)). Within a semiclassical picture we can arrive at the same conclusion if we use the relaxation time approximation for the scattering processes in the Boltzmann equation (Chapter 9).

3.2.2 *Convention for current and voltage*

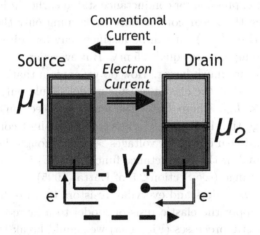

Fig. 3.3 Because an electron carries negative charge, the direction of the electron current is always opposite to that of the conventional current.

Let me briefly comment regarding the direction of the current. As I noted in Chapter 2, because the electronic charge is negative (an unfortunate choice, but we are stuck with it!) the side with the higher voltage has a lower electrochemical potential. Inside the channel, electrons flow from the higher to the lower electrochemical potential, so that the electron current flows from the source to the drain. The conventional current on the other hand flows from the higher to the lower voltage.

Since our discussions will usually involve electron energy levels and the electrochemical potentials describing their occupation, it is also convenient for us to use the electron current instead of the conventional current. For example, in Fig. 3.3 it seems natural to say that the current flows from the source to the drain and not the other way around. And that is what I will try to do consistently throughout this book. *In short, we will use the current, I, to mean electron current.*

3.3   Conductance of a Long Resistor

If the applied bias is much less than kT, we can use Eq. (2.10) to write from Eq. (3.3)

$$I = V \int_{-\infty}^{+\infty} dE \left(-\frac{\partial f_0}{\partial E} \right) G(E)$$

which yields the expression for conductance stated earlier in Eq. (3.1). Since we are obtaining the linear conductance by keeping only the first term in a Taylor series (Eq. (2.10)), it can be justified only for voltages $V < kT/q$, which at room temperature equals 25 mV. But everyday resistors are linear for voltages that are much larger. How do we explain that?

The answer is that the elastic resistor model should only be applied to a short length $< L_{in}$, where L_{in} is the length an electron travels on the average before getting inelastically scattered. Such short conductors could indeed show non-linear effects for voltages $> kT/q$, though the non-linearity may be quite small if the conductance function $G(E)$ is constant over the relevant energy range (see Section 2.5 of Datta (1995)).

But how do we understand everyday resistors that are linear over several volts? To apply the elastic resistor model to a large conductor with distributed inelastic processes (Fig. 3.4a) we should break it up conceptually into a sequence of elastic resistors (Fig. 3.4b), each much shorter than the physical length L, having a voltage that is only a fraction of the total voltage V. As long as the voltage dropped over a length L_{in} is less than kT/q we expect the current to be linear with voltage. The terminal voltage can be much larger.

Before we move on, let me reiterate a little subtlety in viewing a long resistor (Fig. 3.4a) as elastic resistors in series (Fig. 3.4b). We will see in the next chapter that the resistance of an individual section has the form

$$R = R_B \left(1 + \frac{L}{\lambda} \right) \quad \rightarrow \text{ same as Eq. (1.8)} \tag{3.4}$$

and we will see in Part 3, that the length independent part R_B represents an interface resistance associated with the channel-contact interfaces.

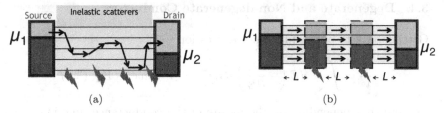

Fig. 3.4 (a) Real conductors have inelastic scatterers distributed throughout the channel. (b) A hypothetical series of elastic resistors as an approximation to a real resistor with distributed inelastic scattering as shown in (a).

Now, the structure in Fig. 3.4b has too many conceptual interfaces that are not present in the real structure of Fig. 3.4a. Each of these interfaces introduces an interface resistance as shown in Fig. 3.5 and these have to be subtracted out. But this is straightforward to do, once we understand the nature and origin of the interface resistance. For example, the resistance of the real structure in Fig. 3.4a of length $3L$ is approximately given by

$$R = R_B \left(1 + \frac{3L}{\lambda}\right) \quad \text{and NOT by} \quad R = 3R_B \left(1 + \frac{L}{\lambda}\right). \tag{3.5}$$

In general our approach is to consider just a single section, exclude the interface resistances to obtain the length dependent part of the resistance $(R_B L/\lambda)$ from which we deduce bulk properties like the conductivity.

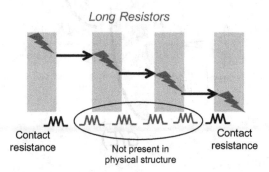

Fig. 3.5 A long resistor can be viewed as a series of ideal elastic resistors. However, we have to exclude the resistance due to all the conceptual interfaces that we introduce which are not present in the physical structure.

3.4   Degenerate and Non-degenerate Conductors

Getting back to our conductance expression

$$\frac{I}{V} = \int_{-\infty}^{+\infty} dE \left(-\frac{\partial f_0}{\partial E} \right) G(E) \quad \rightarrow \quad \text{same as Eq. (3.1)}$$

we note that depending on the nature of the conductance function $G(E)$ and the thermal broadening function we can identify two distinct limits (Fig. 3.6).

The first is case A where the conductance function $G(E)$ is nearly constant over the width of the broadening function. We could then pull $G(E)$ out of the integral in Eq. (3.1) to write

$$\frac{I}{V} \approx G(E = \mu_0) \int_{-\infty}^{+\infty} dE \left(-\frac{\partial f_0}{\partial E} \right) = G(E = \mu_0). \qquad (3.6)$$

This relation suggests an operational definition for the conductance function $G(E)$: *it is the conductance measured at low temperatures for a channel with its electrochemical potential μ_0 located at E.* The actual conductance is obtained by averaging $G(E)$ over a range of energies using $-\partial f_0/\partial E$ as a weighting function. Case A is a good example of the so-called degenerate conductors.

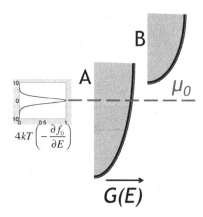

Fig. 3.6   Degenerate (A) and non-degenerate (B) limits.

The other extreme is the non-degenerate conductor shown in case B where the electrochemical potential is located at an energy many kT's below the energy range where the conductance function is non-zero. As a result over the energy range of interest where $G(E)$ is non-zero, we have

$$x \equiv \frac{E - \mu_0}{kT} \gg 1$$

and it is common to approximate the Fermi function with the Boltzmann function

$$\frac{1}{1 + e^x} \approx e^{-x}$$

so that $\quad \dfrac{I}{V} = \displaystyle\int_{-\infty}^{+\infty} \frac{dE}{kT} \, G(E) \, e^{-(E-\mu_0)/kT}.$

This non-degenerate limit is commonly used in the semiconductor literature though the actual situation is often intermediate between degenerate and non-degenerate limits.

We will generally use the expression

$$G = \frac{q^2 D}{2t}$$

with the understanding that the quantities D and t are evaluated at $E = \mu_0$ at low temperatures. Depending on the nature of $G(E)$ and the location of μ_0, we may need to average $G(E)$ over a range of energies using as a "weighting function" as prescribed by Eq. (3.1).

Chapter 4

Ballistic and Diffusive Transport

Related video lectures available at course website, Unit 1: L1.5, L1.6, L1.7 and L1.8.

We saw in the last chapter that the resistance of an elastic resistor can be written as

$$G = \frac{q^2 D}{2t} \quad \rightarrow \text{ same as Eq. (3.2).}$$

In this chapter, I will first argue that the transit time t across a resistor of length L for diffusive transport with a mean free path can be related to the time t_B for ballistic transport by the relation (Section 4.1)

$$t = t_B \left(1 + \frac{L}{\lambda} \right). \tag{4.1}$$

Combining with Eq. (3.2) we obtain

$$G = \frac{G_B \lambda}{L + \lambda} \tag{4.2}$$

$$\text{where } G_B = \frac{q^2 D}{2 t_B}. \tag{4.3}$$

We could rewrite Eq. (4.2) as

$$G = \frac{\sigma A}{L + \lambda} \tag{4.4a}$$

$$\text{where } \sigma A = G_B \lambda. \tag{4.4b}$$

So far we have only talked about three dimensional resistors with a large cross-sectional area A. Many experiments involve two-dimensional resistors

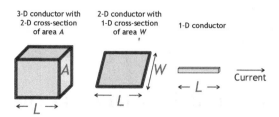

Fig. 4.1   3D, 2D and 1D conductors.

whose cross-section is effectively one-dimensional with a width W, so that the appropriate equations have the form

$$G = \frac{\sigma W}{L + \lambda}$$

where $\sigma W = G_B \lambda.$

Finally we have one-dimensional conductors for which

$$G = \frac{\sigma}{L + \lambda}$$

where $\sigma = G_B \lambda.$

We could collect all these results and write them compactly in the form

$$G = \frac{\sigma}{L + \lambda}\{1, W, A\} \tag{4.5a}$$

where $\sigma = G_B \lambda \left\{1, \dfrac{1}{W}, \dfrac{1}{A}\right\}.$ (4.5b)

The three items in parenthesis correspond to 1D, 2D and 3D conductors. Note that the conductivity has different dimensions in 1D, 2D and 3D, while both G_B and λ have the same dimensions, namely Siemens (S) and meter (m) respectively.

Note that Eq. (4.5) is different from the standard Ohm's law

$$G = \frac{\sigma}{L}\{1, W, A\}$$

which predicts that the resistance will approach zero (conductance will become infinitely large) as the length L is reduced to zero. Of course no one expects it to become zero, but the common belief is that it will approach a value determined by the interface resistance which can be made arbitrarily small with improved contacting technology.

What is now well established experimentally is that even with the best possible contacts, there is a minimum interface resistance determined by the properties of the channel, independent of the contact. The modified Ohm's law in Eq. (4.5) reflects this fact: even a channel of zero length with perfect contacts has a resistance equal to that of a hypothetical channel of length λ. But what does it mean to talk about the mean free path of a channel of zero length? The answer is that neither σ nor λ mean anything for a short conductor, but the ballistic conductance

$$G_B = \frac{\sigma}{\lambda}\{1, W, A\}$$

represents a basic material parameter whose significance has become clear in the light of modern experiments (see Section 4.2).

The ballistic conductance is proportional to the number of channels, $M(E)$ available for conduction, which is proportional to, but not the same as, the density of states, $D(E)$. The concept of density of states has been with us since the earliest days of solid state physics. By contrast, the number of channels (or transverse modes) $M(E)$ is a more recent concept whose significance was appreciated only after the seminal experiments in the 1980s on ballistic conductors showing conductance quantization.

4.1 Transit Times

Consider how the two quantities in

$$G = \frac{q^2 D}{2t}$$

namely the density of states, D and the transfer time t scale with channel dimensions for large conductors. The first of these is relatively easy to see since we expect the number of states to be additive. A channel twice as big should have twice as many states, so that the density of states $D(E)$ for large conductors should be proportional to the volume (AL).

Regarding the transfer time, t, broadly speaking there are two transport regimes:

Ballistic regime: Transfer time $t \sim L$
Diffusive regime: Transfer time $t \sim L^2$.

Consequently the *ballistic conductance is proportional to the area* (note that $D \sim AL$ as discussed above), *but independent of the length.* This "non-ohmic" behavior has indeed been observed in short conductors. It is only diffusive conductors that show the "ohmic" behavior $G \sim A/L$.

These two regimes can be understood as follows.

Ballistic Transport

Source →→→→→ Drain

$\longrightarrow z$

In the ballistic regime electrons travel straight from the source to the drain "like a bullet," taking a time

$$t_B = \frac{L}{\overline{u}} \quad \text{where} \quad \overline{u} = \langle |v_z| \rangle \tag{4.6}$$

is the average velocity of the electrons in the z-direction.

But conductors are typically not short enough for electrons to travel "like bullets." Instead they stumble along, getting scattered randomly by various defects along the way taking much longer than the ballistic time in Eq. (4.6).

Diffusive Transport

Source ⤳⤨⤳ Drain

$\longrightarrow z$

We could write

$$t = \frac{L}{\overline{u}} + \frac{L^2}{2\overline{D}} \tag{4.7}$$

viewing it as a sort of "polynomial expansion" of the transfer time t in powers of L. We could then argue that the lowest term in this expansion must equal the ballistic limit, while the highest term should equal the diffusive limit well-known from the theory of random walks. This theory (see for example, Berg, 1993) identifies the coefficient $\overline{D}$ as the diffusion constant

$$\overline{D} = \langle v_z^2 \tau \rangle$$

τ being the mean free time.

Some readers may not find this "polynomial expansion" completely satisfactory. But this approach has the advantage of getting us to the new Ohm's law (Eq. (4.5)) very quickly using simple algebra. In Chapters 8 and 9 we will obtain this result more directly from the Boltzmann equation.

Getting back to Eq. (4.7), we use Eq. (4.6) to rewrite it in the form

$$t = t_B \left(1 + \frac{L\bar{u}}{2\overline{D}}\right)$$

which agrees with Eq. (4.1) if the mean free path is given by

$$\lambda = \frac{2\overline{D}}{\bar{u}}.$$

In defining the two constants $\overline{D}$ and $\bar{u}$ we have used the symbol $\langle\rangle$ to denote an average over the angular distribution of velocities which yields a different numerical factor depending on the dimensionality of the conductor (see Appendix B). For $d = \{1, 2, 3\}$ dimensions

$$\bar{u} \equiv \langle|v_z|\rangle = v(E) \left\{1, \frac{2}{\pi}, \frac{1}{2}\right\} \tag{4.8}$$

$$\overline{D} \equiv \langle v_z^2 \tau \rangle = v^2(E)\tau(E) \left\{1, \frac{1}{2}, \frac{1}{3}\right\} \tag{4.9}$$

$$\lambda = \frac{2\overline{D}}{\bar{u}} = v(E)\tau(E) \left\{2, \frac{\pi}{2}, \frac{4}{3}\right\}. \tag{4.10}$$

Note that *our definition of the mean free path includes a dimension-dependent numerical factor over and above the standard value of* $v\tau$.

Couldn't we simply use the standard definition? We could, but then the new Ohm's law would not simply involve replacing L with L plus λ. Instead it would involve L plus a dimension-dependent factor times λ. Instead we have chosen to absorb this factor into the definition of λ.

Interestingly, even in one dimensional conductors the factor is not one, but two. This is because the mean free time after which an electron gets scattered. Assuming the scattering to be isotropic, only half the scattering events will result in an electron traveling towards the drain to head towards the source. The mean free time for backscattering is thus, making the mean free path $2v\tau$ rather than $v\tau$.

4.2   Channels for Conduction

Next we obtain an expression for the *ballistic conductance* by combining Eq. (4.3) with Eq. (4.6) to obtain

$$G_B = \frac{q^2 D \bar{u}}{2L}$$

and then make use of Eq. (4.8) to write

$$G_B = \frac{q^2 D v}{2L} \left\{ 1, \frac{2}{\pi}, \frac{1}{2} \right\} \tag{4.11}$$

Eq. (4.11) tells us that the ballistic conductance depends on D/L, the density of states per unit length.

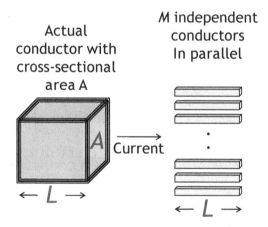

Fig. 4.2   Channels of conduction: a key concept.

Since D is proportional to the volume, the ballistic conductance is expected to be proportional to the cross-sectional area A in 3D conductors (or the width W in 2D conductors). This was experimentally observed in metals in 1969 and is known as the *Sharvin resistance*. Numerous experiments since the 1980s have shown that for small conductors, the ballistic conductance does not go down linearly with the area A. Rather it goes down in integer multiples of the *conductance quantum*

$$G_B = \underbrace{\frac{q^2}{h}}_{\sim 40\mu S} \times \underbrace{M}_{\text{integer}} . \tag{4.12}$$

How can we understand this relation and what does the integer M represent? This result cannot come out of our elementary treatment of electrons in classical particle-like terms, since it involves Planck's constant h. Some input from quantum mechanics is clearly essential and this will come in Chapter 6 when we evaluate $D(E)$. For the moment we note that heuristically Eq. (4.12) suggests that we visualize the real conductor as M independent channels in parallel whose conductances add up to give Eq. (4.11) for the ballistic conductance.

This suggests that we use Eqs. (4.12) and (4.11) to define a quantity $M(E)$ (floor(x) denotes the largest integer less than or equal to x)

$$M = \text{floor}\left(\frac{hDv}{2L} \left\{ 1, \frac{2}{\pi}, \frac{1}{2} \right\} \right) \tag{4.13}$$

which provides a measure of the number of conducting channels. In Chapter 6 we will use a simple model that incorporates the wave nature of electrons to show that for a one-dimensional channel the quantity M indeed equals one showing that it has only one channel, while for two- and three-dimensional conductors the quantity M represents the number of de Broglie wavelengths that fit into the cross-section, like the modes of a waveguide.

Chapter 5

Conductance from Fluctuation

5.1 Introduction

In this chapter we will digress a little to connect our conductance formula

$$G = \frac{q^2 D}{2t} \quad \rightarrow \text{ same as Eq. (3.2)}$$

to the very powerful fluctuation-dissipation theorem widely used in discussing linear transport coefficients, like the conductivity. In our discussion we have stressed the non-equilibrium nature of the problem of current flow requiring contacts with different electrochemical potentials (see Fig. 2.4).

Just as heat flow is driven by a difference in temperatures, current flow is driven by a difference in electrochemical potentials. Our basic current expression (see Eqs. (3.2) and (3.3))

$$I = q \int_{-\infty}^{+\infty} dE \, \frac{D(E)}{2t(E)} \, (f_1(E) - f_2(E)) \tag{5.1}$$

is applicable to arbitrary voltages but so far we have focused largely on the low bias approximation (see Eqs. (3.1) and (3.2))

$$G_0 = q^2 \int_{-\infty}^{+\infty} dE \left(-\frac{\partial f_0}{\partial E} \right) \frac{D(E)}{2t(E)} \tag{5.2}$$

Although we have obtained this result from the general non-equilibrium expression, it is interesting to note that the low bias conductance is really an equilibrium property. Indeed there is a fundamental theorem relating the low bias conductance for small voltages to the fluctuations in the current that occur at equilibrium when no voltage is applied. Consider a conductor with no applied voltage (see Fig. 5.1) so that both source and drain have the same electrochemical potential μ_0. There is of course no net current without an applied voltage, but even at equilibrium, every once in a while, an electron crosses over from source to drain and on the average an equal number crosses over the other way from the drain to the source, so that

$$\langle I(t_0) \rangle_{eq} = 0$$

where the angular brackets $\langle \rangle$ denote either an "ensemble average" over many identical conductors or more straightforwardly a time average over the time t_0.

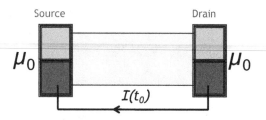

Fig. 5.1   At equilibrium both contacts have the same electrochemical potential μ_0. No net current flows, but there are equal currents I_0 from source to drain and back.

However, if we calculate the current correlation

$$C_I = \int_{-\infty}^{+\infty} d\tau \langle I(t_0 + \tau) I(t_0) \rangle_{eq} \tag{5.3}$$

we get a non-zero value even at equilibrium, and the fluctuation-dissipation (F-D) theorem relates this quantity to the low bias conductance:

$$G_0 = \frac{C_I}{2kT} = \frac{1}{2kT} \int_{-\infty}^{+\infty} d\tau \langle I(t_0 + \tau) I(t_0) \rangle_{eq}. \tag{5.4}$$

This is a very powerful result because it allows one to calculate the conductance by evaluating the current correlations using the methods of equilibrium statistical mechanics, which are in general more well-developed than the methods of non-equilibrium statistical mechanics. Indeed before the advent of mesoscopic physics in the late 1980s, the Kubo formula based on the F-D theorem was the only approach used to model quantum transport. The Kubo formula in principle applies to large conductors with inelastic scattering, though in practice it may be difficult to evaluate the effect of complicated inelastic processes on the current correlation.

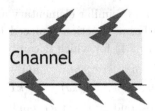

The usual approach is to evaluate transport in long conductors with a high frequency alternating voltage, for which electrons can slosh back and forth without ever encountering the contacts. One could then obtain the zero frequency conductivity by letting the sample size L tend to infinity *before* letting the frequency tend to zero (see for example, Chapter 5 of Doniach and Sondheimer (1974)). However, this approach is limited to linear response. In this book (part B) we will stress the Non-Equilibrium Green's Function (NEGF) method for quantum transport, which allows us to address the non-equilibrium problem head on for quantum transport, just as the Boltzmann equation (BTE) does for semiclassical transport.

In this chapter my purpose is primarily to connect our discussion to this very powerful and widely used approach. We will look at the effect of contacts on the current correlations in an elastic resistor and show that applied to an elastic resistor, the F-D theorem (Eq. (5.4)) does lead to our old result (Eq. (3.2)) from Chapter 3.

Interestingly, our elementary arguments in Chapter 3 lead to a conductance proportional to

$$\frac{f_1 - f_2}{\mu_1 - \mu_2} \simeq -\frac{\partial f_0}{\partial E}$$

while the current correlations in the F-D theorem lead to

$$\frac{f_0(1 - f_0)}{kT}$$

with the $1 - f_0$ factor arising from the exclusion principle. The physical arguments are very different, but their equivalence is ensured by the identity

$$-\frac{\partial f_0}{\partial E} = \frac{f_0(E)(1 - f_0(E))}{kT} \tag{5.5}$$

which can be verified with a little algebra, starting from the definition of the Fermi function (Eq. (2.2)).

For phonons (Chapter 14) similar elementary arguments lead to a similar expression with the Fermi function replaced by the Bose function, n (Eq. (14.5) and Section 15.5.1) for which it can be shown that

$$\frac{n_1 - n_2}{\hbar\omega} \simeq -\frac{\partial n}{\partial(\hbar\omega)} = \frac{n(1+n)}{kT}. \tag{5.6}$$

Agreement with the corresponding F-D theorem in this case requires a $1+n$ factor instead of the $1-f$ factor for electrons. This is of course the well-known phenomenon of stimulated emission for Bose particles. We will talk a little more about Fermi and Bose functions in Chapter 15.

5.2 Current Fluctuations in an Elastic Resistor

5.2.1 *One-level resistor*

Consider first a one-level resistor connected to two contacts with the same electrochemical potential μ_0 and hence the same Fermi function $f_0(E)$ (see Fig. 5.2).

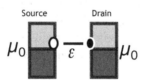

Fig. 5.2   At equilibrium with the same electrochemical potential in both contacts, there is no net current. But there are random pulses of current as electrons cross over in either direction.

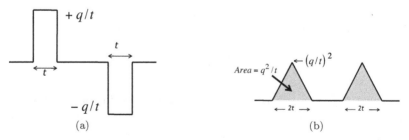

Fig. 5.3   (a) Random pulses of current at equilibrium with the same electrochemical potential. (b) Current correlation function.

There are random positive and negative pulses of current as electrons cross over from the source to the drain and from the drain to the source respectively. The average positive current is equal to the average negative current, which we call the equilibrium current I_0 and write it in terms of the transit time t

$$I_0 = \frac{q}{t} f_0(\varepsilon)(1 - f_0(\varepsilon)) \tag{5.7}$$

where the factor $f_0(\epsilon)(1 - f_0(\epsilon))$ is the probability that an electron will be present at the source ready to transfer to the drain but no electron will be present at the drain ready to transfer back. The correlation is obtained by treating the transfer of each electron from the source to the drain as an independent stochastic process. The integrand in Eq. (5.3) then looks like a sequence of triangular pulses as shown each having an area of q^2/t, so that

$$C_I = 2\frac{q^2}{t} f_0(\varepsilon)(1 - f_0(\varepsilon)) \tag{5.8}$$

where the additional factor of 2 comes from the fact that I_0 only counts the positive pulses, while both positive and negative pulses contribute additively to C_I.

5.2.2 Multi-level resistor

To generalize our one-level results from Eq. (5.7) to an elastic resistor with an arbitrary density of states, $D(E)$ we note that in an energy range dE there are $D(E)\,dE$ states so that

$$I = q \int_{-\infty}^{+\infty} dE \, \frac{D(E)}{2t(E)} f_0(E)(1 - f_0(E)) \tag{5.9a}$$

$$C_I = 2q^2 \int_{-\infty}^{+\infty} dE \, \frac{D(E)}{2t(E)} f_0(E)(1 - f_0(E)) \tag{5.9b}$$

assuming that the fluctuations in different energy ranges can simply be added, as we are doing by integrating over energy.

Note that $C_I = 2qI_0$ suggesting that the fluctuation is like the shot noise due to the equilibrium currents I_0 flowing in either direction. Making use of Eq. (5.4) we have for the conductance

$$G_0 = \frac{C_I}{2kT} = q^2 \int_{-\infty}^{+\infty} dE \, \frac{f_0(E)(1 - f_0(E))}{kT} \frac{D(E)}{2t(E)} \tag{5.10}$$

which can be reduced to our old expression (see Eqs. (3.1) and (3.2)) making use of the identity stated earlier in Eq. (5.5).

Before moving on, let me note that there is at present an extensive body of work on subtle correlation effects in elastic resistors (see for example, Splettstoesser *et al.* 2010), some of which have been experimentally observed. But the theory of noise even for an elastic resistor is more intricate than the theory for the average current that we will focus on in this book.

PART 2

Simple Model for Density of States

Chapter 6

Energy Band Model

Related video lecture available at course website, Unit 2: L2.10.

6.1 Introduction

A common expression for conductivity is the Drude formula relating the conductivity to the electron density n, the effective mass m and the mean free time τ

$$\sigma \equiv \frac{1}{\rho} = \frac{q^2 n \tau}{m}.$$

$$(6.1)$$

This expression is very well-known since even freshman physics texts start by deriving it (see Section 2.5.1). It also leads to the widely used concept of mobility

$$\tilde{\mu} = \frac{q\tau}{m}$$

$$(6.2)$$

$$\text{with} \quad \sigma = qn\tilde{\mu}.$$

$$(6.3)$$

On the other hand, Eq. (4.5b) expresses the conductivity as a product of the ballistic conductance G_B and the mean free path λ

$$\sigma = G_B \lambda \left\{ 1, \ \frac{1}{W}, \ \frac{1}{A} \right\} \quad \text{(same as Eq. (4.5b))}.$$

$$(6.4)$$

This expression can be rewritten, using Eq. (4.3) for G_B, Eq. (4.6) for t_B and Eq. (4.10) for λ, as a product of the density of states D and the diffusion coefficient $\overline{D}$ (see Eq. (4.9))

$$\sigma(E) = q^2 \, \overline{D} \, \frac{D}{L} \left\{ 1, \ \frac{1}{W}, \ \frac{1}{A} \right\}$$

$$(6.5)$$

59

Eq. (6.5) is a standard result referred to as the degenerate Einstein relation but it is not as well-known as the Drude formula. Most people remember Eq. (6.1) and not Eq. (6.5). Our objective in this chapter is to relate the two.

Note that, like the conductance (see Eq. (3.1)), these expressions for the energy-dependent conductivity also have to be averaged over an energy range of a few kT's around $E = \mu_0$, using the thermal broadening function,

$$\sigma_0 = \int_{-\infty}^{+\infty} dE \left(-\frac{\partial f_0}{\partial E} \right) \sigma(E). \tag{6.6}$$

It is this averaged conductivity that should be compared to the Drude conductivity in Eq. (6.1). But for degenerate conductors (see Eq. (3.6)) the averaged conductivity is approximately equal to the conductivity at an energy $E = \mu_0$:

$$\sigma_0 \approx \sigma(E = \mu_0) \tag{6.7}$$

and so we can compare $\sigma(E = \mu_0)$ from Eq. (6.5) with Eq. (6.1).

The point we wish to stress is that while Eq. (6.1) is often very useful, it is a result of limited validity that can be obtained from Eq. (6.5) by making suitable approximations based on a specific model. But when these approximations are not appropriate, we can still use Eq. (6.5) which is *far more generally applicable*.

For example, Eq. (6.5) gives sensible answers even for materials like graphene whose non-parabolic bands make the meaning of mass somewhat unclear, causing considerable confusion when using Eq. (6.1). In general we should really use Eq. (6.5), and not Eq. (6.1), to shape our thinking about conductivity.

There is a fundamental difference between Eqs. (6.5) and (6.1). The averaging implied in Eq. (6.6) makes the conductivity a "Fermi surface property", that is one that depends only on the energy levels close to $E = \mu_0$. By contrast, Eq. (6.1) depends on the total electron density n integrated over all energy. But this dependence on the total number is true only in a limited sense.

Experts know that "n" only represents the density of "free" electrons and have an instinctive feeling for what it means to be free. They know that there are p-type semiconductors which conduct better when they have fewer electrons, but in that case they know that n should be interpreted to mean the number of "holes". For beginners, all this appears confusing and much of this confusion can be avoided by using Eq. (6.5) instead of Eq. (6.1).

Interestingly, Eq. (6.5) was used in a seminal paper to obtain Eq. (3.2)

$$G(E) = \frac{q^2 D(E)}{2t(E)} \quad \text{(same as Eq. (3.2))}. \tag{6.8}$$

Equation (1) of Thouless (1977) is essentially the same as this equation with minor differences in definitions. What we have done is to use the concept of an elastic resistor to first obtain Eq. (3.2) from elementary arguments, and then used it to obtain Eq. (6.5).

Equation (6.5) stresses that the essential factor determining the conductivity is the density of states around $E = \mu_0$. Materials are known to have conductivities ranging over many orders of magnitude from glass to copper. And the basic fact remains that they all have approximately the same number of electrons. Glass is an insulator but not because it is lacking in electrons. It is an insulator because it has a very low density of states or number of modes around $E = \mu_0$.

So when does Eq. (6.5) reduce to (6.1)? Answer: if the electrons are described by a "single band effective mass model" as I will try to show in this chapter. So far we have kept our discussion general in terms of the density of states, $D(E)$ and the velocity, $v(E)$ without adopting any specific models. These concepts are generally applicable even to amorphous materials and molecular conductors. A vast amount of literature both in condensed matter physics and in solid state devices, however, is devoted to crystalline solids with a periodic arrangement of atoms because of the major role they have played from both basic and applied points of view.

For such materials, energy levels over a limited range of energies are described by an $E(p)$ relation and we will show in this chapter that irrespective of the specific $E(p)$ relation, at any energy E the density of states $D(E)$, velocity $v(E)$ and momentum $p(E)$ are related to the total number of states $N(E)$ with energy less than E by the relation (d: number of dimensions)

$$D(E)v(E)p(E) = N(E) \cdot d. \tag{6.9}$$

We can combine this relation with Eq. (6.5) and make use of Eq. (4.9), to write

$$\sigma(E) = \frac{q^2 \tau(E)\nu(E)}{p(E)} \left\{ \frac{N(E)}{L}, \quad \frac{N(E)}{WL}, \quad \frac{N(E)}{AL} \right\}. \tag{6.10}$$

Equation (6.10) indeed looks like Drude expression (Eq. (6.1)) if we identify (1) the mass as

$$m(E) = \frac{p(E)}{v(E)} \tag{6.11}$$

which is independent of energy for parabolic $E(p)$ relations, but can in general be energy-dependent, and (2) the quantity in parenthesis

$$\left\{ \frac{N(E)}{L}, \quad \frac{N(E)}{WL}, \quad \frac{N(E)}{AL} \right\}$$

as the electron density, n per unit length, area and volume in 1D, 2D and 3D respectively. At low temperatures, this is easy to justify since the energy averaging in Eq. (6.6) amounts to looking at the value at $E = \mu_0$ and $N(E)$ at $E = \mu_0$ represents the total number of electrons (Fig. 6.1).

At non-zero temperatures one needs a longer discussion which we will get into later in the chapter. Indeed as will see, some subtleties are involved even at zero temperature when dealing with differently shaped density of states.

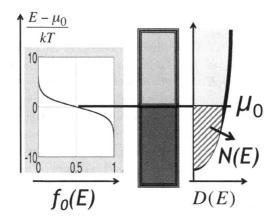

Fig. 6.1 Equilibrium Fermi function $f_0(E)$, density of states $D(E)$ and integrated density of states $N(E)$. This is an example of an n-type conductor as opposed to a p-type conductor shown later in Fig. 6.2.

Note, however, that the key to reducing our conductivity expression (Eq. (6.5)) to the Drude-like expression (Eq. (6.10)) is Eq. (6.9) which is an interesting relation for it relates $D(E)$, $v(E)$ and $p(E)$ at a given energy E, to the total number of states $N(E)$ obtained by integrating $D(E)$

$$N(E) = \int_{-\infty}^{E} dE\, D(E).$$

How can the integrated value of $D(E)$ be uniquely related to the value of quantities like $D(E)$, $v(E)$ and $p(E)$ at a single energy? The answer is that this relation holds only as long as the energy levels are given by a single $E(p)$ relation. It may not hold in an energy range with multiple bands of energies or in an amorphous solid not described by an $E(p)$ relation. Equation (6.1) is then not equivalent to Eq. (6.5), and *it is* **Eq. (6.5)** *that can be trusted*.

With that long introduction let us now look at how single bands described by an $E(p)$ relation leads to Eq. (6.9) and helps us connect our conductivity expression (Eq. (6.5)) to the Drude formula (Eq. (6.1)). This will also lead to a different interpretation of the quantity $M(E)$ introduced in the last chapter, that will help understand why it is an integer representing the number of channels.

6.2 $E(p)$ or $E(k)$ Relation

Related video lecture available at course website, Unit 2: L2.2.

The general principle for calculating $D(E)$ is to start from the Schrödinger equation treating the electron as a wave confined to the solid. Confined waves (like a guitar string) have resonant "frequencies" and these are basically the allowed energy levels. By counting the number of energy levels in a range E to $E + dE$, we obtain the density of states $D(E)$.

Although the principle is simple, a first principles implementation is fairly complicated since one needs to start from a Schrödinger equation including an appropriate potential that the electrons feel inside the solid not only due to the nuclei but also due to the other electrons.

One of the seminal concepts in solid state physics is the realization that in crystalline solids electrons behave as if they are in vacuum, but with an effective mass different from their natural mass, so that the energy-momentum relation can be written as

$$E(p) = E_c + \frac{p^2}{2m} \qquad (6.12)$$

where E_c is a constant.

The momentum p is equated to $\hbar k$, providing the link between the energy-momentum relation $E(p)$ associated with the particle viewpoint and the dispersion relation $E(k)$ associated with the wave viewpoint. Here we will write everything in terms of p, but they are easily translated using the relation $p = \hbar k$.

Equation (6.12) is generally referred to as a parabolic dispersion relation and is commonly used in a wide variety of materials from metals like copper

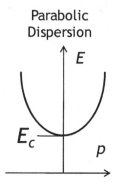

to semiconductors like silicon, because it often approximates the actual $E(p)$ relation fairly well in the energy range of interest. But it is by no means the only possibility. Graphene, a material of great current interest, is described by a linear relation:

$$E = E_c + v_0 p \qquad (6.13)$$

where v_0 is a constant. Note that p denotes the magnitude of the momentum and we will assume that the $E(p)$ relation is isotropic, which means that it is the same regardless of which direction the momentum vector points.

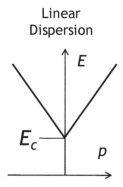

For any given isotropic $E(p)$ relation, the velocity points in the same direction as the momentum, while its magnitude is given by

$$v \equiv \frac{dE}{dp}. \qquad (6.14)$$

This is a general relation applicable to arbitrary energy-momentum relations for classical particles. On the other hand, in wave mechanics it is justified as the group velocity for a given dispersion relation $E(k)$. Note that Eq. (6.14) yields an energy-independent mass $m = p/v$, only for a parabolic $E(p)$ relation (Eq. (6.12)) and not for the linear relation in Eq. (6.13).

6.3 Counting States

Related video lecture available at course website, Unit 2: L2.3.

One great advantage of this principle is that it reduces the complicated problem of electron waves in a solid to that of waves in vacuum, where the allowed energy levels can be determined the same way we find the resonant frequencies of a guitar string: simply by requiring that an integer number of wavelengths fit into the solid. Noting that the de Broglie principle relates the electron wavelength to the Planck's constant divided by its momentum, h/p, we can write

$$\frac{L}{h/p} = Integer \Rightarrow p = Integer \times \left(\frac{h}{L}\right) \tag{6.15}$$

where L is the length of the box. This means that the allowed states are uniformly distributed in p with each state occupying a "space" of

$$\triangle p = \frac{h}{L}. \tag{6.16}$$

Let us define a function $N(p)$ that tells us the total number of states that have a momentum less than a given value p. In *one dimension* this function is written down by dividing the total range of $2p$ (from $-p$ to $+p$) by the spacing h/L:

$$N(p) = \frac{2p}{h/L} = 2L\left(\frac{p}{h}\right) \rightarrow 1D.$$

In two dimensions we divide the area of a circle of radius p by the spacing $h/L \times h/W$, L and W being the dimensions of the two dimensional box.

$$N(p) = \frac{\pi p^2}{(h/L)(h/W)} = \pi WL\left(\frac{p}{h}\right)^2 \rightarrow 2D.$$

In three dimensions we divide the volume of a sphere of radius p by the spacing $h/L \times h/W_1 \times h/W_2$, L, W_1 and W_2 being the dimensions of the three dimensional box. Writing $A = W_1 \times W_2$ we have

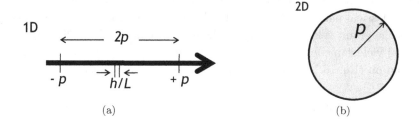

$$N(p) = \frac{(4\pi/3)p^3}{(h/L)(h^2/A)} = \frac{4\pi}{3} AL \left(\frac{p}{h}\right)^3 \to 3D.$$

We can combine all three results into a single expression for $d = \{1, 2, 3\}$ dimensions:

$$N(p) = \left\{ 2\frac{L}{h/p}, \quad \pi\frac{LW}{(h/p)^2}, \quad \frac{4\pi}{3}\frac{LA}{(h/p)^3} \right\}. \tag{6.17}$$

6.3.1 Density of states, D(E)

Related video lecture available at course website, Unit 2: L2.4.

We could use a given $E(p)$ relation to turn this function $N(p)$ into a function of energy $N(E)$ that tells us the total number of states with energy less than E, which must equal the density of states $D(E)$ *integrated* up to an energy E, so that $D(E)$ can be obtained from the derivative of $N(E)$:

$$N(E) = \int_{-\infty}^{E} dE' D(E') \to D(E) = \frac{dN}{dE}.$$

Making use of Eqs. (6.17) and (4.8),

$$D(E) = \frac{dN}{dp}\frac{dp}{dE} \to D(E)v(E) = \frac{dN}{dp} = d \cdot \frac{N}{p}$$

leading to the relation stated earlier

$$D(E)v(E)p(E) = N(E) \cdot d \quad \text{(same as Eq. (6.9))}.$$

Note that *this identity is independent of the actual $E(p)$ relation.*

6.4 Number of Modes

Related video lecture available at course website, Unit 2: L2.5.

I noted in Chapter 4 that the ballistic conductance is given by

$$G_B = \underbrace{\frac{q^2}{h}}_{\sim 40\mu S} \times \underbrace{M}_{\text{integer}} \quad \text{(same as Eq. (4.12))}$$

and that experimentally M is found to be an integer in low dimensional conductors at low temperatures. Based on this observation we defined

$$M = \text{floor}\left(\frac{hDv}{2L}\left\{1, \quad \frac{2}{\pi}, \quad \frac{1}{2}\right\}\right) \quad \text{(same as Eq. (4.13))}$$

but there was no justification for choosing integer values for M instead of letting it be a continuous variable as the simple semiclassical theory suggested. Using the $E(p)$ relations discussed in this chapter we will now show that we can interpret $M(p)$ in a very different way that helps justify its integer nature. First we make use of Eq. (6.9) to rewrite Eq. (4.13) in the form (dropping the floor function for the moment)

$$M = \frac{hN}{2Lp}\left\{1, \quad \frac{4}{\pi}, \quad \frac{3}{2}\right\} \tag{6.18}$$

where $N(p)$ is the total number of states with a momentum that is less than p and we have seen that it is equal to the number of wavelengths that fit into the solid. Making use of Eq. (6.17) for $N(p)$, we obtain

$$M(p) = \left\{1, \quad 2\frac{W}{h/p}, \quad \pi\frac{A}{(h/p)^2}\right\}. \tag{6.19}$$

This result is independent of the actual $E(p)$ relation, since we have not made use of any specific relationship. We can now understand why we should modify Eq. (6.19) to write

$$M(p) = \text{floor}\left\{1, \quad 2\frac{W}{h/p}, \quad \pi\frac{A}{(h/p)^2}\right\} \tag{6.20}$$

where floor(x) represents the largest integer less than or equal to x. Just as $N(p)$ tells us the number of wavelengths that fits into the volume, $M(p)$ *tells us the number that fits into the cross-section.*

If we evaluate our expressions for $N(p)$ and $M(p)$ for a given sample we will in general get a fractional number. However, since these quantities represent the number of states, we would expect them to be integers and if we obtain say 201.59, we should take the lower integer 201.

This point is commonly ignored in large conductors at high temperatures, where experiments do not show this quantization because of the energy averaging over $\mu_0 \pm 2kT$ associated with experimental measurements. For example, if over this energy range, $M(E)$ varies from say 201.59 to 311.67, then it seems acceptable to ignore the fact that it really varies from 201 to 311.

But in small structures where one or more dimensions is small enough to fit only a few wavelengths the integer nature of M is observable and shows up in the quantization of the ballistic conductance. We should then use Eq. (6.20) and not (6.19).

6.4.1 *Degeneracy factor*

One little "detail" that we need to take into account when comparing to experiment is the degeneracy factor "g" denoting the number of equivalent states given by the product of the number of spins and the number of valleys (which we will discuss in Part B). All these g channels conduct in parallel so that ballistic conductors have a resistance of

$$\frac{h}{q^2 M} \approx \frac{25 \text{ k}\Omega}{M} \frac{1}{g}.$$

So the resistance of a 1D ballistic conductor is approximately equal to 25 kΩ divided by g. This has indeed been observed experimentally. Most metals and semiconductors like GaAs have $g = 2$ due to the two spins, and the 1D ballistic resistance ~ 12.5 kΩ. But carbon nanotubes have two valleys as well making $g = 4$ and exhibit a ballistic resistance ~ 6.25 kΩ.

Another "detail" to note is that for two- and three-dimensional conductors, Eq. (6.17) is not quite right, because it is based on the heuristic idea of counting modes by counting the number of wavelengths that fit into the solid (see Eq. (6.15)). Mathematically it can be justified only if we assume periodic boundary conditions, that is if we assume that the cross-section is in the form of a ring rather than a flat sheet for a 2D conductor. For a 3D conductor it is hard to visualize what periodic boundary conditions might look like though it is easy to impose it mathematically as we have been doing.

Ring-shaped
conductor

(a)

Flat
Conductor

(b)

Most real conductors do not come in the form of rings, yet periodic boundary conditions are widely used because it is mathematically convenient and people believe that the actual boundary conditions do not really matter. But this is true only if the cross-section is large. For small area conductors the actual boundary conditions do matter and we cannot use Eq. (6.15).

Interestingly a conductor of great current interest has actually been studied in both forms: a ring-shaped form called a carbon nanotube and a flat form called graphene. If the circumference or width is tens of nanometers they have much the same properties, but if it is a few nanometers their properties are observably different including their ballistic resistances.

6.5 Electron Density, n

Related video lecture available at course website, Unit 2: 'L2.6.

As we mentioned in the introduction, a key quantity appearing in the familiar Drude formula is the electron density, n. In this Section we would like to establish that the total number of electrons can be identified with the function $N(E)$ discussed in Section 6.2. This is easy to see at low temperatures where the energy averaging in Eq. (6.6) amounts to looking at the value at a single energy $E = \mu_0$. Since $N(E)$ represents the total number of states available below E, and the total number of electrons at low temperatures equals the number of states below μ_0, it seems clear that $N(E)$ can be identified with the number of electrons. It then follows that the electron density can be written as

$$n(E) = N(E) \left\{ \frac{1}{L}, \quad \frac{1}{WL}, \quad \frac{1}{AL} \right\}. \tag{6.21}$$

But does this work at non-zero temperatures? We will argue below that the answer is yes.

6.5.1 n-type conductors

Will we get the correct electron density if we energy average $n(E)$ from Eq. (6.21) following the prescription in Eq. (6.6)? It is straightforward to check that the answer is yes, if we carry out the integral "by parts" to yield

$$\int_{-\infty}^{+\infty} dE \left(-\frac{\partial f_0}{\partial E}\right) N(E) = [-N(E) \, f_0(E)]_{-\infty}^{+\infty} + \int_{-\infty}^{+\infty} dE \left(\frac{dN}{dE}\right) f_0(E)$$

$$= [0 - 0] + \int_{-\infty}^{+\infty} dE \, D(E) f_0(E)$$

$$= Total \ Number \ of \ Electrons$$

since $dED(E)f_0(E)$ tells us the number of electrons in the energy range from E to $E+dE$. When integrated it gives us the total number of electrons.

6.5.2 p-type conductors

An interesting subtlety is involved when we consider a p-type conductor for which the $E(p)$ relation extends downwards, say something like

$$E(p) = E_v - \frac{p^2}{2m}.$$

Instead of

$$N(E) = \int_{-\infty}^{E} dE' \, D(E')$$

we now have (see Fig. 6.2)

$$N(E) = \int_{E}^{+\infty} dE' \, D(E') \rightarrow D(E) = -\frac{dN}{dE}.$$

This is because we defined the function $N(E)$ from $N(p)$ which represents the total number of states with momenta less than p, which means energies greater than E for a p-type dispersion relation. Now if we carry out the integration by parts as before

$$\int_{-\infty}^{+\infty} dE \left(-\frac{\partial f_0}{\partial E}\right) N(E) = [-N(E) \, f_0(E)]_{-\infty}^{+\infty} + \int_{-\infty}^{+\infty} dE \left(\frac{dN}{dE}\right) f_0(E)$$

we run into a problem because the first term does not vanish at the lower limit where both $N(E)$ and $f_0(E)$ are both non-zero. We can get around

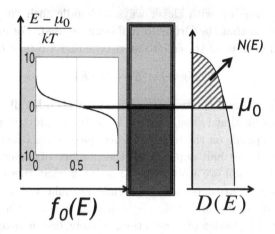

Fig. 6.2 Equilibrium Fermi function $f_0(E)$, density of states $D(E)$ and integrated density of states $N(E)$: p-type conductor.

this problem by writing the derivative in terms of $1 - f_0$ instead of f_0:

$$\int_{-\infty}^{+\infty} dE \left(\frac{\partial(1 - f_0)}{\partial E} \right) N(E)$$

$$= [-N(E)(1 - f_0(E))]_{-\infty}^{+\infty} + \int_{-\infty}^{+\infty} dE \, \frac{dN}{dE} \, [1 - f_0(E)]$$

$$= [0 - 0] + \int_{-\infty}^{+\infty} dE \, D(E) \, [1 - f_0(E)]$$

$$= Total \; Number \; of \; \text{``holes''}.$$

What this means is that with p-type conductors we can use the Drude formula Eq. (6.1)

$$\sigma = q^2 n \tau / m$$

but the n now represents the density of empty states or holes. A larger n really means fewer electrons.

6.5.3 *"Double-ended" density of states*

How would we count n for a density of states $D(E)$ that extends in both directions as shown in Fig. 6.3 (left panel). This is representative of graphene, a material of great interest (recognized by the 2010 Nobel prize in physics), whose $E(p)$ relation is commonly approximated by

$$E = \pm v_0 \, p.$$

People usually come up with clever ways to handle such "double-ended" density of states so that the Drude formula can be used. For example they divide the total density of states into an n-type and a p-type component

$$D(E) = D_n(E) + D_p(E)$$

as shown in Fig. 6.3 and the two components are then handled separately, using a prescription that is less than obvious: the conductivity due to the upper half D_n depends on the number of occupied states (electrons), while that due to the lower half depends on the number of unoccupied states (holes). But the point we would like to stress is that there is really no particular reason to insist on using a Drude formula and keep inventing clever ways to make it work. One might just as well use Eq. (6.5) which reflects the correct physics of conduction, namely *that it takes place in a narrow band of energies around μ_0.*

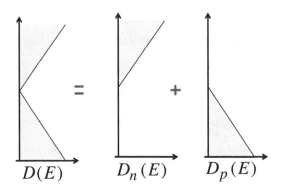

Fig. 6.3   A "double-ended" density of states can be visualized as a sum of an "n-type component" and a "p-type component."

6.6   Conductivity versus n

Related video lecture available at course website, Unit 2: L2.7.

From our new perspective, the conductivity can be written by combining Eq. (4.5b) with Eq. (4.12)

$$\sigma(E) \quad = \quad \frac{q^2}{h} M \lambda \left\{ 1, \ \frac{1}{W}, \ \frac{1}{A} \right\} \tag{6.22}$$

while the Drude formula (Eq. (6.10)) with $m = p/v$ would express it as

$$\sigma(E) = q^2 \frac{v\tau}{p} \frac{N}{L} \left\{ 1, \quad \frac{1}{W}, \quad \frac{1}{A} \right\}. \tag{6.23}$$

The two expressions can be shown to be equivalent making use of Eq. (6.18) for the number of modes and Eq. (4.10) for the mean free path.

Experimental conductivity measurements are often performed as a function of the electron density and the common expectation based on the Drude formula is that conductivity should be proportional to the electron density and any non-linearity must be a consequence of the energy-dependence of the mean free time. Is this generally true? Our expression in Eq. (6.22) suggests that we view the conductivity as the product of the ballistic resistance (or number of modes) and the mean free path. From Eqs. (6.19) and (6.17) we have

$$M \sim p^{d-1} \quad \text{while} \quad n \sim p^d$$

$$\text{so that} \quad G_B \sim M \sim n^{(d-1)/d} \to \sqrt{n} \text{ for } d = 2. \tag{6.24}$$

Short ballistic samples of graphene indeed show this dependence discussed above (see for example, Bolotin *et al.* (2008)). What about the conductivity of long diffusive samples?

The conductivity is a product of G_B and the mean free path, the latter being a product of the velocity and the mean free time. Since graphene has a linear energy-momentum relation (Eq. (6.13)), the velocity is a constant independent of energy, and if the mean free time were constant too, the conductivity too would be $\sqrt{n}$.

In practice, however, the mean free time for specific scattering mechanisms $\tau(E) \sim \sqrt{n}$, so that the conductivity often ends up being proportional to the electron density, n (see Torres *et al.* 2014 for a thorough discussion). But the point to note is that the energy dependence of the mass, $m(E)$ and the mean free time $\tau(E)$ happen to cancel out accidentally, to give $\sigma \sim n$.

Before we move on I should again mention the little "detail" that I mentioned at the end of Section 6.4. This is the degeneracy factor g which denotes the number of equivalent states. For example all non-magnetic materials have two spin states with identical energies, which would make $g = 2$. Certain materials also have equivalent "valleys" having identical energy momenta relations so that the N we calculate for one valley has to be multiplied by g when relating to the experimentally measured electron densities. For graphene, $g = 2 \times 2 = 4$.

All our discussion applies to a single spin and valley for which the conductance and the electron density are each $1/g$ times the actual, so that Eq. (6.24) gets modified to

$$\frac{G_B}{g} \sim \left(\frac{n}{g}\right)^{(d-1)/d} \quad \to \quad G_B \sim g^{1/d}\, n^{(d-1)/d}. \tag{6.25}$$

Chapter 7

The Nanotransistor

Related video lecture available at course website, Unit 2: L2.9.

Our "field-less" approach to conductivity comes as a surprise to many since it is commonly believed that currents are driven by electric fields. However, we hasten to add that the field can and does play an important role once we go beyond low bias and our purpose in this chapter is to discuss the role of the electrostatic potential and the corresponding electric field on the current-voltage characteristics beyond low bias.

To illustrate these issues, I will use the nanotransistor, an important device that is at the heart of microelectronics. As we noted at the outset the nanotransistor is essentially a voltage-controlled resistor whose length has shrunk over the years and is now down to a few hundred atoms. But as any expert will tell you, it is not just the low bias resistance, but the entire shape of the current-voltage characteristics of a nanotransistor that determines its utility. And this shape is controlled largely by its electrostatics, making it a perfect example for our purpose.

I should add, however, that this chapter does not do justice to the nanotransistor as a device. This will be discussed in a separate volume in this series written by Lundstrom, whose model is widely used in the field and forms the basis of our discussion here. We will simply use the nanotransistor to illustrate the role of electrostatics in determining current flow.

We have seen that the elastic transport model leads to the current formula

$$I = \frac{1}{q} \int_{-\infty}^{+\infty} dE \ G(E) \ (f_1(E) - f_2(E)) \quad \text{(see Eq. (3.3))}.$$

In this chapter, I will use the nanotransistor to illustrate a few issues that need to be considered at high bias, some of which can be modeled with a simple extension of Eq. (3.3)

$$I = \frac{1}{q} \int_{-\infty}^{+\infty} dE \ G(E - U) \ (f_1(E) - f_2(E)) \tag{7.1}$$

to include an appropriate choice of the potential U in the channel which is treated as a single point. We call this the *point channel model* to distinguish it from the standard and more elaborate extended channel model which we will introduce at the end of the chapter.

7.1 Current-voltage Relation

The nanotransistor is a three-terminal device (Fig. 7.1), though ideally no current should flow at the gate terminal whose role is just to control the current. In other words, the current-drain voltage, I-V_D, characteristics are controlled by the gate voltage, V_G (see Fig. 7.2). The low bias current and conductance can be understood based on the principles we have already discussed. But currents at high V_D involve important new principles.

The basic principle underlying an FET is straightforward (see Fig. 7.3). A positive gate voltage V_G changes the potential in the channel, lowering all the states down in energy, which can be included by setting $U = -qV_G$ in Eq. (7.1).

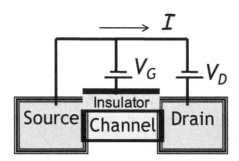

Fig. 7.1 Sketch of a field effect transistor (FET): channel length, L; transverse width, W (perpendicular to page).

For an n-type conductor this increases the number of available states in the energy window of interest around μ_1 and μ_2 as shown. Of course for a p-type conductor (see Fig. 6.2) the reverse would be true leading to a

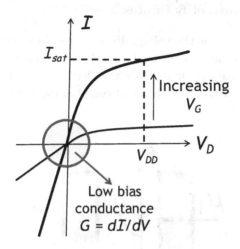

Fig. 7.2 Typical current-voltage, I-V_D characteristic and its variation with V_G for an FET with an n-type channel of the type shown in Fig. 7.1 built on an *insulating* substrate so that the drain voltage V_D can be made either positive or negative as shown. This may not be possible in FETs built on semiconducting substrates and standard textbooks normally do not show negative V_D for n-MOSFETs.

complementary FET (see Fig. 1.2) whose conductance variation is just the opposite of what we are discussing. But we will focus here on n-type FETs.

We will not discuss the low bias conductance since these involve no new principles. Instead we will focus on the current at high bias, specifically on why the current-voltage, I-V_D characteristic is (1) non-linear, and (2) "rectifying", that is different for positive and negative V_D.

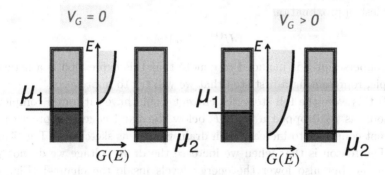

Fig. 7.3 A positive gate voltage V_G increases the current in an FET by moving the states down in energy.

7.2 Why the Current Saturates

Figure 7.2 shows that as the voltage V_D is increased the current does not continue to increase linearly. Instead it levels off tending to saturate. Why? The reason seems easy enough. Once the electrochemical potential in the drain has been lowered below the band edge the current does not increase any more (Fig. 7.4).

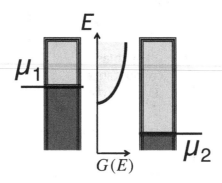

Fig. 7.4 The current saturates once μ_2 drops below the band-edge.

The saturation current can be written from Eq. (7.1)

$$I_{sat} = \frac{1}{q} \int_{-\infty}^{+\infty} dE \; G(E - U) \; f_1(E) \qquad (7.2)$$

by dropping the second term $f_2(E)$ assuming μ_2 is low enough that $f_2(E)$ is zero for all energies where the conductance function is non-zero. In the simplest approximation

$$U^{(1)} = -q\,V_G.$$

The superscript 1 is included to denote that this expression is a little too simple, representing a first step that we will try to improve.

If this were the full story the current would have saturated completely as soon as μ_2 dropped a few kT below the band edge. In practice the current continues to increase with drain voltage as sketched in Fig. 7.6.

The reason is that when we increase the drain voltage we do not just lower μ_2, but also lower the energy levels inside the channel (Fig. 7.5) similar to the way a gate voltage would. The result is that the current keeps increasing as the conductance function $G(E)$ slides down in energy

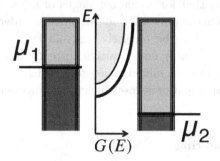

Fig. 7.5 The current does not saturate completely because the states in the channel are also lowered by the drain voltage.

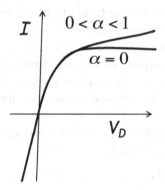

Fig. 7.6 Current in an FET would saturate perfectly if the channel potential were unaffected by the drain voltage.

by a fraction α (< 1) of the drain voltage V_D, which we could include in our model by choosing

$$U^{(2)} \rightarrow \quad U_L \quad \equiv \quad \alpha(-q\,V_D) + \beta(-q\,V_G). \qquad (7.3)$$

Indeed the challenge of designing a good transistor is to make α as small as possible so that the channel potential is hardly affected by the drain voltage. If α were zero the current would saturate perfectly as shown in Fig. 7.6 and that is really the ideal: a device whose current is determined entirely by V_G and not at all by V_D or in technical terms, a high transconductance but low output conductance. For reasons we will not go into, this makes designing circuits much easier.

To ensure that V_G has far greater control over the channel than V_D it is necessary to make the insulator thickness a small fraction of the channel length. This means that for a channel length of a few hundred atoms we need an insulator that is only a few atoms thick in order to ensure a small α. This thickness has to be precisely controlled since thinner insulators would lead to unacceptably large leakage currents. We mentioned earlier that today's laptops have a billion transistors. What is even more amazing is that each has an insulator whose thickness is precisely controlled down to a few atoms!

7.3 Role of Charging

There is a second effect that leads to an increase in the saturation current over what we get using Eq. (7.3) in Eq. (7.1). Under bias, the occupation of the channel states is less than what it is at equilibrium. This is because at equilibrium both contacts are trying to fill up the channel states, while under bias only the source is trying to fill up the states while the drain is trying to empty it. Since there are fewer electrons in the channel, it tends to become positively charged and this will lower the states in the channel as shown in Fig. 7.5, resulting in an increase in the current, even if the electrostatics is perfect ($\alpha = 0$) and the drain voltage has no effect on the channel.

This effect can be captured within the point channel model (Eq. (7.1)) by writing the channel potential as

$$(A) \qquad U = U_L + U_0(N - N_0) \qquad (7.4)$$

where U_L is given by our previous expression in Eq. (7.3). The extra term represents the change in the channel potential due to the change in the number of electrons in the channel, N under non-equilibrium conditions relative to the equilibrium number N_0, U_0 being the change in the channel potential energy per electron. To use Eq. (7.4), we need expressions for N_0 and N. N_0 is the equilibrium number of channel electrons, which can be calculated simply by filling up the density of states, $D(E)$ according to the equilibrium Fermi function $f_0(E)$.

$$(B1) \qquad N_0 = \int_{-\infty}^{+\infty} dE \ D(E - U) \ f_0(E) \qquad (7.5)$$

while the number of electrons, N in the channel under non-equilibrium conditions is given by

$$(B2) \qquad N = \int_{-\infty}^{+\infty} dE \ D(E - U) \ \frac{f_1(E) \ + \ f_2(E)}{2} \qquad (7.6)$$

assuming that the channel is "equally" connected to both contacts. Note that the calculation is now a little more intricate than what it would be if U_0 were zero. We now have to obtain a solution for U and N that satisfy both Eqs. (7.4) and (7.6) simultaneously through an iterative procedure as shown schematically in Fig. 7.7.

Once a self-consistent U has been obtained, the current is calculated from Eq. (7.1), or an equivalent version that is sometimes more convenient numerically and conceptually.

$$(C) \qquad I = \frac{1}{q} \int_{-\infty}^{+\infty} dE \; G(E) \; (f_1(E + U) - f_2(E + U)). \qquad (7.7)$$

This simple point channel model often provides a good description of the *I-V* characteristics as discussed in Rahman *et al.* (2003).

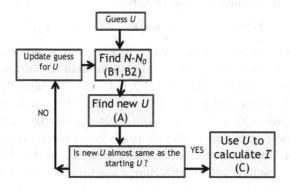

Fig. 7.7 Self-consistent procedure for calculating the channel potential U in point channel model.

Figure 7.9 shows the current versus voltage characteristic calculated numerically (MATLAB code at end of chapter) assuming a 2D channel with a parabolic dispersion relation for which the density of states is given by (L: length, W: width, ϑ: unit step function)

$$D(E) = g \frac{m L W}{2\pi \hbar^2} \vartheta(E - E_c). \qquad (7.8)$$

The numerical results are obtained using $g = 2$, $m = 0.2 \times 9.1 \times 10^{-31}$ Kg, $\beta = 1$, $\alpha = 0$ and $U_0 = 0$ or $U_0 = \infty$ as indicated, with $L = 1$ μm, $W = 1$ μm assuming ballistic transport, so that

$$G(E) = \frac{q^2}{h} M(E).$$

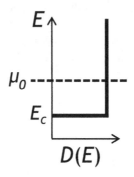

Fig. 7.8 Density of states $D(E)$ given in Eq. (7.8).

$M(E)$ being the number of modes given by

$$M(E) = g\frac{2W}{h}\sqrt{2m(E - E_c)}\,\vartheta(E - E_c). \qquad (7.9)$$

The current-voltage characteristics in Fig. 7.9 has two distinct parts, the initial linear increase followed by a saturation of the current. Although these results were obtained numerically, both the slope and the saturation current can be calculated analytically, especially if we make the low temperature approximation that the Fermi functions change abruptly from 1 to 0 as the energy E crosses the electrochemical potential μ. Indeed we used a kT of 5 meV instead of the usual 25 meV, so that the numerical . results would compare better with simple low temperature estimates.

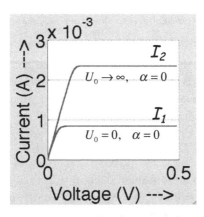

Fig. 7.9   Current-voltage characteristics calculated numerically using the self-consistent point channel model shown in Fig. 7.7. MATLAB code included at end of chapter.

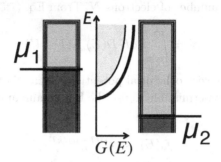

There are two key points we wanted to illustrate with this example. Firstly, the initial slope of the current-voltage characteristics is unaffected by the charging energy. This slope defines the low bias conductance that we have been discussing till we came to this chapter. The fact that it remains unaffected is reassuring and justifies our not bringing up the role of electrostatics earlier.

Secondly, the saturation current is strongly affected by the electrostatics and changes by a factor of ~ 2.8 from a model with zero charging energy to one with a very large charging energy. This is because of the reason mentioned at the beginning of this section. With $U_0 = 0$, the channel states remain fixed and the number of electrons N is equal to $N_0/2$, since $f_1 = 1$ and $f_2 = 0$ in the energy range of interest. With very large U_0, to avoid $U_0(N - N_0)$ becoming excessive, N needs to be almost equal to N_0 even though the states are only half-filled. This requires the states to move down as sketched with a corresponding increase in the current.

7.3.1 *Quantum capacitance*

Related video lecture available at course website, Unit 2: L2.8.

We have generally focused on the shape of the current-voltage characteristics obtained when a voltage is applied between the source and the drain, which is a non-equilibrium problem. Let us take a brief detour to talk about an equilibrium problem where charging can have a major effect. Suppose the source and drain are held at the same potential while the gate voltage V_G is changed. How much does the number of electrons (N) change?

A positive gate voltage lowers the density of states $D(E)$ resulting in an increase in the number of electrons N. From Eq. (7.6)

$$N = \int_{-\infty}^{+\infty} dE\, D(E)\, f_0(E + U). \qquad (7.10)$$

Assuming that the electrochemical potential is located outside the band as shown, it is usually permissible to use the Boltzmann approximation to the Fermi function

$$f_0(E) \approx e^{-(E-\mu)/kT}$$

so that $\quad N = e^{-U/kT} \times \int_{-\infty}^{+\infty} dE\, D(E)\, f_0(E).$

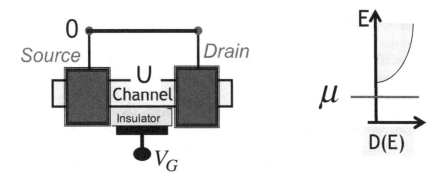

To change the number of electrons by a factor of 10, we need a change in the channel potential U by

$$kT \ln(10) \approx 60 \text{ meV}.$$

To change the channel potential U by 60 meV, we need a gate voltage of at least 60 mV, leading to an oft-quoted result that one needs a gate voltage of at least 60 mV for every decade of change in N.

This discussion, however, is relevant only when the number of electrons is relatively small. Once the electrochemical potential gets close to the band, it is important to include the charging effect.

A positive gate voltage tries to lower the density of states as shown by the solid line, which increases the number of electrons, but that in turn causes the states to float up as indicated by the dashed line.

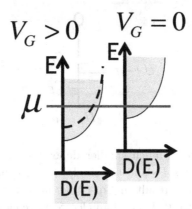

To determine the resulting N in general, one has to include this charging effect. We can divide the problem into two parts, (1) the change in N due to a change in the channel potential U, and (2) the change in U due to a change in the gate voltage V_G:

$$C = q\frac{dN}{dV_G} = q\frac{dN}{(dU/-q)} \times \frac{dU}{d(-qV_G)}. \qquad (7.11)$$

The first term is the quantum capacitance which can be related to the density of states starting from Eq. (7.8) as before (but not invoking the Boltzmann approximation)

$$C_Q = q\frac{dN}{(dU/-q)}$$

$$= q^2 \int_{-\infty}^{+\infty} dE\, D(E) \left(-\frac{\partial f_0}{\partial E}\right) = q^2 D_0 \qquad (7.12)$$

where D_0 represents the averaged value of the density of states $D(E)$ around $E + U = \mu$, that is, around $E = \mu - U$.

To evaluate the second term we write from Eq. (7.4), using Eq. (7.12),

$$\frac{dU}{d(V_G)} = \frac{dU_L}{d(V_G)} + U_0\frac{dN}{d(V_G)} = \frac{dU_L}{d(V_G)} - U_0 D_0\frac{dU}{d(V_G)}.$$

Hence

$$\frac{dU}{d(-qV_G)} = \frac{1}{1 + U_0 D_0} \times \frac{dU_L}{d(-qV_G)}. \qquad (7.13)$$

The second term represents the change in the potential U_L due to a gate voltage which ideally could approach one. But the actual change in the

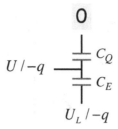

channel potential can be much smaller depending on the product of the average density of states D_0 and the single electron charging energy U_0.

We can visualize the result in Eq. (7.13) in terms of the quantum capacitance defined in Eq. (7.12) and an electrostatic capacitance C_E related to the charging energy

$$U_0 = \frac{q^2}{C_E}$$

$$\text{so that} \quad U_0 D_0 = \frac{C_Q}{C_E}. \qquad (7.14)$$

The electrostatic capacitance reflects the charging energy related to the increase in the potential of the channel q/C_E when a single electron is added to it.

In this book we have been talking about the density of states D_0 from the outset, but we have talked very little about the charging effects U_0. This is because the low bias conductance of a homogeneous structure is ordinarily not affected by it, but we would like to stress that charging is generally an integral part of device analysis, both in equilibrium and out of equilibrium.

7.4 "Rectifier" Based on Electrostatics

Let us now look at an example that can be handled using the point channel model just discussed though it does not illustrate any issues affecting the design of nanotransistors.

I have chosen this example to illustrate a fundamental point that is often not appreciated, namely that an otherwise symmetric structure could exhibit asymmetric current-voltage characteristics (which we are loosely calling a "rectifier"). In other words, we could have

$$I(+V_D) \neq I(-V_D)$$

for a symmetric structure, simply because of *electrostatic asymmetry.*

Consider a nanotransistor having perfect electrostatics represented by $\alpha = 0$ (Eq. (7.3)), connected (a) in the standard configuration (Fig. 7.10a) and (b) with the gate left floating (Fig. 7.10b). The basic device is assumed physically symmetric, so that one could not tell the difference between the source and drain contacts. This is usually true of real transistors, but that is not important, since we are only trying to make a conceptual point.

The configuration in (a) has electrostatic asymmetry, since the gate is held at a fixed potential with respect to the source, but not with respect to the drain. But configuration (b) is symmetric in this respect too, since the gate floats to a potential halfway between the source and the drain. Figure 7.11 shows the current-voltage characteristics calculated using the model summarized in Fig. 7.7 (MATLAB code at end of chapter), for each of the structures shown in Figs. 7.10(a) and (b). The parameters are the same as those used for the example shown in Fig. 7.9, except that the equilibrium electrochemical potential is located exactly at the bottom of the band as shown in Fig. 7.10: $\mu_0 = E_c$.

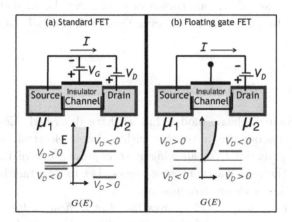

Fig. 7.10   (a) Standard FET assuming perfect electrostatics. (b) Floating gate FET.

The standard FET connection corresponds to $\alpha = 0$ assuming perfect electrostatics, while the same physical structure in the floating gate connection corresponds to $\alpha = 0.5$. The former gives a rectifying characteristic, while the latter gives a linear characteristic, often called "ohmic". The point is that it is not necessary to design an asymmetric channel to get asymmetric I-V characteristics. Even the simplest symmetric channel can exhibit

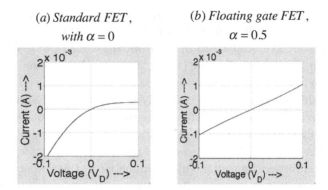

Fig. 7.11 Current-voltage characteristics obtained from the point channel model corresponding to the configurations shown in Fig. 7.10. MATLAB code included at end of chapter.

non-symmetric $I(V_D)$ characteristic if the electrostatics is asymmetric.

Note also that the linear conductance given by the slope dI/dV around $V = 0$ is unaffected by our choice of α and can be predicted without any reference to the electrostatics, even though the overall shape obviously cannot.

7.5 Extended Channel Model

The point channel elastic model that we have described (Eqs. (7.1) and (7.4)) integrates our elastic resistor with a simple electrostatic model for the channel potential U/q, allowing it to capture some of the high bias physics that the pure elastic resistor misses. Let us now talk briefly about some of the things we are still missing.

The point channel model *ignores the electric field in the channel* and assumes that the density of states $D(E)$ stays the same from source to drain. In the real structure, however, the electric field lowers the states at the drain end relative to the source as sketched here. Doesn't this change the current?

For an elastic resistor one could argue that the additional states with the slanted (rather than horizontal) shading are not really available for conduction since (in an elastic resistor) every energy represents an independent energy channel and can only conduct if it connects all the way from the source to the drain.

But even for an elastic resistor there should be an increase in current because at a given energy E, the number of modes at the drain end is larger than the number of modes at the source end. This is because the number of modes at an energy E depends on how far this energy is from the bottom of the band determined by $U(z)$ (see Eq. (7.9)) which is lower at the drain than at the source.

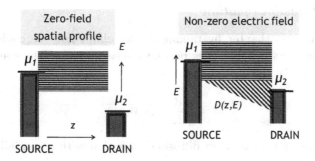

Fig. 7.12   An electric field in the channel causes the density of states $D(E)$ to increase from the source to the drain.

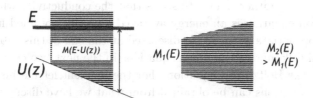

Fig. 7.13   The number of channels $M(E)$ is larger at the drain end than at the source because of the lower $U(z)$.

The structure almost looks as if it were "wider" at the drain than at the source. For a ballistic conductor this makes no difference since the conductance function cannot exceed the maximum set by the "narrowest" point. But for a conductor that is many mean free paths long, the broadening at the drain could increase the conductance relative to that of an un-broadened channel.

In general

$$\frac{q^2}{h}\frac{M_1\lambda}{L+\lambda} \leq G(E) \leq \frac{q^2}{h}M_1. \tag{7.15}$$

This effect is not very important for near ballistic elastic channels, since the minimum and maximum values of the conductance function in Eq. (7.15) are then essentially equal. But it can lead to a significant increase in the current for diffusive channels as the drain voltage V_D is increased. A semi-analytical expression describing this effect is given in Appendix C.

7.5.1   Diffusion equation

Let me end this chapter by talking briefly about how we could develop a quantitative model that goes beyond point channels. The standard approach is to take the continuity equation

$$\frac{d}{dz}I = 0 \tag{7.16}$$

and combine it with a "drift-diffusion" equation with a spatially varying conductivity:

$$\frac{I}{A} = -\frac{\sigma_0(z)}{q}\frac{d\mu}{dz}. \tag{7.17}$$

We are using σ_0 rather than σ to stress that the conductivity which enters the diffusion equation is an energy-averaged quantity obtained from $\sigma(E)$ and the appropriate averaging is discussed at the end of this subsection.

We will talk more about these equations in the next two chapters when we discuss the Boltzmann equation. For the moment let me just indicate how these equations can be obtained from what we have discussed so far.

Equation (7.16) is easy to see. If we have a current of 25 electrons per second entering a section of the conductor and only 10 electrons per second leaving it, then the number of electrons will be building up in this section at the rate of $25 - 10 = 15$ per second. That is a transient condition, not a steady-state one. Under steady-state conditions the current has to be the same at all points along the z-axis as required by Eq. (7.16).

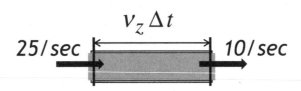

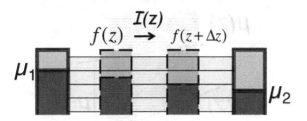

To obtain Eq. (7.17) we view a long conductor as a series of elastic resistors as discussed in Section 3.3. Using Eq. (3.3) we can write the current $I(z)$ in a section of the conductor as

$$I(z) = \frac{1}{q} \int_{-\infty}^{+\infty} dE \, G(E) \, (f(z, E) - f(z + \triangle z, E)).$$

From Eq. (4.5a) we could write

$$\frac{1}{G(E)} = \rho \, \frac{\triangle z + \lambda}{A}$$

but the point to note is that part of this resistance represents the interface resistance, which should not be included since there are no actual interfaces except at the very ends. Omitting the interface resistance we can write (Note: $\sigma = 1/\rho$)

$$G(E) = \frac{\sigma A}{\triangle z}.$$

Combining this with our usual linear expansion for small potential differences from Eq. (2.10)

$$(f(z, E) - f(z + \triangle z, E)) \equiv \left(-\frac{\partial f_0}{\partial E}\right) (\mu(z, E) - \mu(z + \triangle z, E))$$

and defining the conductivity as the thermal average of $\sigma(E)$ (Eq. (6.6)), we obtain

$$I(z) = \frac{1}{q} \frac{\sigma_0 A}{\triangle z} (\mu(z) - \mu(z + \triangle z)).$$

Letting $\triangle \to 0$, we obtain Eq. (7.17).

What do we use for the conductivity, σ_0? Our old expression

$$\sigma_0 = \int_{-\infty}^{+\infty} dE \, \sigma(E) \left(-\frac{\partial f}{\partial E}\right)_{E=\mu_0} \qquad \text{(same as Eq. (6.6))}$$

involved an energy average of over an energy window of a few kT around $E = \mu_0$.

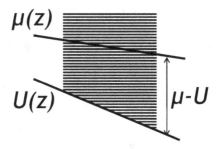

The spatially varying $U(z)$ shifts the available energy states in energy, so that one now has to look at the energy window around $E = \mu(z) - U(z)$ suggesting that we replace Eq. (6.6) with

$$\sigma_0 = \int_{-\infty}^{+\infty} dE\, \sigma(E) \left(-\frac{\partial f}{\partial E} \right)_{E=\mu(z)-U(z)}. \tag{7.18}$$

7.5.2 *Charging: Self-consistent solution*

Note that to use Eq. (7.18) we have to determine $\mu(z) - U(z)$ from a self-consistent solution the Poisson equation (ϵ: Permittivity, n_0, n: electron density per unit volume at equilibrium and out of equilibrium)

$$(A') \quad \frac{d}{dz}\left(\epsilon \frac{dU}{dz} \right) = q^2 (n - n_0). \tag{7.19}$$

The electron density per unit length entering the Poisson equation is calculated by filling up the density of states (per unit length) shifted by the local potential $U(z)$, according to the local electrochemical potential, so that we can write

$$(B1') \quad n(z) \equiv \int_{-\infty}^{+\infty} dE\, \frac{D(E - U(z))}{L} \frac{1}{1 + \exp\left(\dfrac{E - \mu(z)}{kT} \right)} \tag{7.20a}$$

$$(B2') \quad n_0 = \int_{-\infty}^{+\infty} dE\, \frac{D(E)}{L} \frac{1}{1 + \exp\left(\dfrac{E - \mu_0}{kT} \right)}. \tag{7.20b}$$

Solving Eq. (7.20) self-consistently with the Poisson equation (Eq. (7.19)) is indeed the standard approach to obtaining the correct $\mu(z)$ and $U(z)$, which can then be used to find the current from Eq. (7.17). We could view this procedure as the extended channel version of the point channel model in Fig. 7.7 as shown in Fig. 7.14.

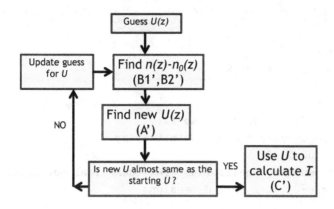

Fig. 7.14 Extended channel version of the point channel model in Fig. 7.7.

Note that this whole approach is based on the assumption of a local electrochemical potential $\mu(z)$ appearing in Eq. (7.20a). In general, electron distributions can deviate so badly from Fermi functions that an electrochemical potential may not be adequate and one needs the full semiclassical formalism based on the Boltzmann Transport Equation (BTE) which we discuss in Chapter 9.

Much progress has been made in this direction but full-fledged BTE-based simulation is time-consuming and the drift-diffusion equation based on the concept of a local potential $\mu(z)$ continues to be the "bread and butter" of device modeling. But the diffusion equation has been around a long time. What are we adding to it? It is the extra interface resistance that changes

$$G(E) \;=\; \frac{\sigma A}{L} \;\text{ to }\; \frac{\sigma A}{L + \lambda}.$$

In the next chapter we will show that this result follows from the diffusion equation if the boundary conditions are modified appropriately and to justify this modified boundary condition requires a deeper discussion of the meaning of the electrochemical potential on a nanometer scale.

7.6 MATLAB Codes for Figs. 7.9 and 7.11

These codes are included here mainly for their pedagogical value. Soft copies are available through our website
https://nanohub.org/groups/lnebook
% Saturation current

```
clear all

% Constants
hbar=1.06e-34;q=1.6e-19;

%Parameters
m=0.2*9.1e-31;g=2;ep=4*8.854e-12;
% mass, degeneracy factor, permittivity
kT=0.005;mu0=0.05;
% Thermal energy, equilibrium electrochemical potential
W=1e-6;L=1e-6;tox=2e-9;% dimensions

% Energy grid
dE=0.00001;E=[0:dE:2];NE=size(E,2);
D=(g*W*L*q*m/2/pi/hbar/hbar).*ones(1,NE);
% Density of states
M=(g*2*W/2/pi/hbar).*sqrt(2*m*q*E);% Number of modes

f0=1./(1+exp((E-mu0)./kT));
% Equilibrium Fermi function
N0=sum(dE*D.*f0);% Equilibrium electron number
U0=0*q*tox/ep/L/W;% Charging energy
alpha=0.05;% Drain induced barrier lowering

% I-V characteristics
ii=1;dV=0.01;for V=1e-3:dV:0.5
UL=-alpha*V;ii
 change=100;U=UL;

% Self-consistent loop
 while change>1e-6
    f1=1./(1+exp((E-mu0+U)./kT));
    f2=1./(1+exp((E-mu0+U+V)./kT));
    f=(f1+f2)./2;
   N=sum(dE*D.*f);% Electron number
   Unew=UL+U0*(N-N0);% Self-consistent potential
     change=abs(U-Unew);
     U=U+0.05*(Unew-U);
 end
```

```
    curr(ii)=(q*q*dE/2/pi/hbar)*sum(M.*(f1-f2));
    volt(ii)=V;Uscf(ii)=U;ns(ii)=N/W/L;ii=ii+1;
end
current=q*g*(4/3)*W*sqrt(2*m*q*mu0)*q*mu0/
4/pi/pi/hbar/hbar;
max(curr)/current

h=plot(volt,curr,'r');
%h=plot(volt,curr,'r+');
%h=plot(volt,Uscf,'r');
%h=plot(volt,ns,'r');
set(h,'linewidth',[3.0])
set(gca,'Fontsize',[40])
xlabel(' Voltage (V_{D}) ---> ');
ylabel(' Current (A) ---> ');


% Electrostatic rectifier
clear all

% Constants
hbar=1.06e-34;q=1.6e-19;

%Parameters
m=0.2*9.1e-31;g=2;ep=4*8.854e-12;
% mass, degeneracy factor, permittivity
kT=0.025;mu0=0*0.05;
% Thermal energy, equilibrium electrochemicalpotential
W=1e-6;L=1e-6;tox=2e-9;% dimensions

% Energy grid
dE=0.0001;E=[0:dE:2];NE=size(E,2);
D=(g*W*L*q*m/2/pi/hbar/hbar).*ones(1,NE);
% Density of states
M=(g*2*W/2/pi/hbar).*sqrt(2*m*q*E);% Number of modes

f0=1./(1+exp((E-mu0)./kT));
% Equilibrium Fermi function
```

```
N0=sum(dE*D.*f0);% Equilibrium electron number
U0=0*q*tox/ep/L/W;% Charging energy
alpha=0;% Drain induced barrier lowering

% I-V characteristics
ii=1;dV=0.01;for V=-0.1:dV:0.1
UL=-alpha*V;ii
 change=100;U=UL;

% Self-consistent loop
 while change>1e-6
    f1=1./(1+exp((E-mu0+U)./kT));
    f2=1./(1+exp((E-mu0+U+V)./kT));
    f=(f1+f2)./2;
   N=sum(dE*D.*f);% Electron number
   Unew=UL+U0*(N-N0);% Self-consistent potential
     change=abs(U-Unew);
     U=U+0.05*(Unew-U);
  end

    curr(ii)=(g*q*q*dE/2/pi/hbar)*sum(M.*(f1-f2));
    volt(ii)=V;Uscf(ii)=U;ns(ii)=N/W/L;ii=ii+1;
end

h=plot(volt,curr,'r');
%h=plot(volt,Uscf,'r');
%h=plot(volt,ns,'r');
set(h,'linewidth',[3.0])
set(gca,'Fontsize',[40])
xlabel(' Voltage (V_{D}) ---> ');
ylabel(' Current (A) ---> ');
```

PART 3

What and Where is the Voltage Drop

Chapter 8

Diffusion Equation for Ballistic Transport

8.1 Introduction

Related video lecture available at course website, Unit 3: L3.2.

The title of this chapter may sound contradictory, like the elastic resistor. Doesn't the diffusion equation describe diffusive transport? How can one use it for ballistic transport? An important idea we are trying to get across with our bottom-up approach is the essential unity of these two regimes of transport and hopefully this chapter will help.

In Chapter 7 we introduced the continuity equation

$$\frac{dI}{dz} = 0 \quad \text{(same as Eq. (7.16))} \tag{8.1}$$

and the diffusion equation

$$\frac{I}{A} = -\frac{\sigma_0}{q}\frac{d\mu}{dz} \quad \text{(same as Eq. (7.17))} \tag{8.2}$$

where σ_0 is the energy-averaged conductivity (Eq. (6.6)). In Chapter 9 we will see how this equation is formally obtained from the Boltzmann equation. For the moment let us talk about what we do with these equations.

The standard approach is to solve Eqs. (8.1) and (8.2) with the boundary conditions

$$\mu(z = 0) = \mu_1 \tag{8.3a}$$

$$\mu(z = L) = \mu_2. \tag{8.3b}$$

99

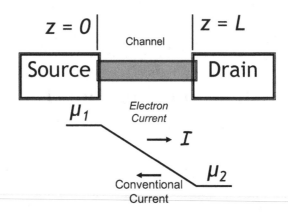

Fig. 8.1   Solution to Eqs. (8.1) and (8.2) with the boundary conditions in Eq. (8.3). Note that we are using I to represent the electron current as explained earlier (see Fig. 3.3).

It is easy to see that the linear solution sketched in Fig. 8.1 meets the boundary conditions in Eq. (8.3) and at the same time satisfies both Eqs. (8.1) and (8.2) since a linear $\mu(z)$ has a constant slope given by

$$\frac{d\mu}{dz} = -\frac{\mu_1 - \mu_2}{L}$$

so that from Eq. (8.2) we have a constant current with $dI/dz = 0$:

$$I = \frac{\sigma_0 A}{q}\frac{\mu_1 - \mu_2}{L}.$$

Note that $\mu_1 - \mu_2 = qV$ (Eq. (2.1)), so that

$$I = \frac{\sigma_0 A}{L}V \tag{8.4}$$

which is the standard expression and not the generalized one we have been discussing

$$I = \frac{\sigma_0 A}{L + \lambda}V \tag{8.5}$$

that includes ballistic channels as well. Can we obtain this result (Eq. (8.5)) from the diffusion equation (Eq. (8.2))?

Many would say that a whole new approach is needed since quantities like the conductivity or the diffusion coefficient mean nothing for a ballistic channel. The central result I wish to establish in this chapter is that we can still use Eq. (8.2) provided we modify the boundary conditions in Eq. (8.3) to reflect the interface resistance that we have been talking about:

$$\mu(z=0) \; = \; \mu_1 - \frac{qIR_B}{2} \tag{8.6a}$$

$$\mu(z=L) \; = \; \mu_2 + \frac{qIR_B}{2} \tag{8.6b}$$

R_B being the inverse of the ballistic conductance G_B discussed earlier (see Eqs. (4.11) and (4.12)):

$$R_B \; = \; \frac{\lambda}{\sigma_0 A} \; = \; \frac{h}{q^2 M}. \tag{8.7}$$

The new boundary conditions in Eqs. (8.6) can be visualized in terms of lumped resistors $R_B/2$ at the interfaces as shown in Fig. 8.2 leading to additional potential drops as shown.

It is straightforward to see that this *new boundary condition* applied to a uniform resistor leads to the new Ohm's law in Eq. (8.5). Since $\mu(z)$

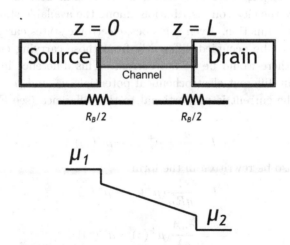

Fig. 8.2 Equations. (8.1) and (8.2) can be used to model both ballistic and diffusive transport provided we modify the boundary conditions in Eq. (8.3) to Eq. (8.6) reflect the two interface resistances, each equal to $R_B/2$.

varies linearly from $z = 0$ to $z = L$, the current is obtained from Eq. (8.2)

$$I = \frac{\sigma_0 A}{q} \frac{\mu(0) - \mu(L)}{L}.$$

Using Eqs. (8.6)

$$I = \frac{\sigma_0 A}{q} \left(\frac{\mu_1 - \mu_2}{L} - \frac{qIR_B}{L} \right).$$

Since, $\sigma_0 A R_B = \lambda$ (Eq. (8.7)),

$$I \left(1 + \frac{\lambda}{L} \right) = \frac{\sigma_0 A}{q} \left(\frac{\mu_1 - \mu_2}{L} \right).$$

Noting that $\mu_1 - \mu_2 = qV$ (Eq. (2.1)) this yields Eq. (8.5).

But how do we justify this new boundary condition (Eqs. (8.6))? It follows from the new Ohm's law (Eq. (8.5)) if we assume that the extra resistance corresponding to $L = 0$ is equally divided between the two interfaces. For a better justification, we need to introduce two different electrochemical potentials μ^+ and μ^- for electrons moving along $+z$ and $-z$ respectively. In previous chapters we talked about *electrochemical potentials inside the contacts* which are large regions that always remain close to equilibrium and hence are described by Fermi functions (see Eqs. (2.7)) with well-defined electrochemical potentials.

By contrast in this chapter we are using $\mu(z)$ to represent quantities *inside the out-of-equilibrium channel*, where it is at best an approximate concept since the electron distribution among the available states need not follow a Fermi function. Even if it does, electronic states carrying current along $+z$ must be occupied differently from those carrying current along $-z$, or else there would be no net current. This difference in occupation is reflected in different electrochemical potentials μ^+ and μ^- and we will show that the current is proportional to the difference (see Eq. (8.23) in Section 8.3)

$$I = \frac{q}{h} M(\mu^+(z) - \mu^-(z)) \tag{8.8}$$

which can also be rewritten in the form

$$I = \frac{1}{qR_B}(\mu^+(z) - \mu^-(z)) \tag{8.9}$$

$$= \frac{\sigma_0 A}{q\lambda}(\mu^+(z) - \mu^-(z)) \tag{8.10}$$

using Eq. (4.4b). The correct boundary conditions for μ^+ and μ^- are

$$\mu^+(z = 0) = \mu_1 \tag{8.11}$$

$$\mu^-(z = L) = \mu_2 \tag{8.12}$$

which can be understood by noting that at $z = 0$ the electrons moving along $+z$ have just emerged from the left contact and hence have the same distribution and electrochemical potential, μ_1. Similarly at $z = L$ the electrons moving along $-z$ have just emerged from the right contact and thus have the same potential μ_2 (Fig. 8.3).

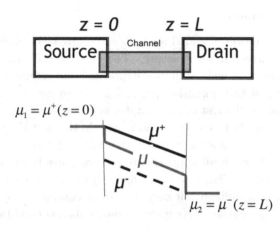

Fig. 8.3   Spatial profile of electrochemical potentials μ^+ and μ^- across a diffusive channel.

In Chapter 9, I will show that the current is related to the potentials μ^+ and μ^- by an equation

$$I = -\frac{\sigma_0 A}{q}\left(\frac{d\mu^+}{dz}\right) = -\frac{\sigma_0 A}{q}\left(\frac{d\mu^-}{dz}\right) \tag{8.13}$$

that looks just like the diffusion equation (Eq. (8.2)) which applies to the average potential:

$$\mu(z) = \frac{\mu^+(z) + \mu^-(z)}{2}. \tag{8.14}$$

Equation (8.13) can be solved with the boundary conditions in Eqs. (8.11) and (8.12) to obtain the plot shown in Fig. 8.3 for μ^+ and μ^- and their average indeed looks like Fig. 8.2 for μ with its discontinuities at the ends. However, it is not necessary to abandon the traditional diffusion equation (Eq. (8.2)) in favor of the new diffusion equation (Eq. (8.13)). We can obtain the same results simply by modifying the boundary conditions for $\mu(z)$ as follows:

$$\mu(z = 0) = \left(\frac{\mu^+ + \mu^-}{2}\right)_{(z=0)} = \left(\mu^+ - \frac{\mu^+ - \mu^-}{2}\right)_{(z=0)} = \mu_1 - \left(\frac{qIR_B}{2}\right)$$

making use of Eqs. (8.9) and (8.11). Similarly using Eqs. (8.9) and (8.12)

$$\mu(z = L) = \left(\mu^- + \frac{\mu^+ - \mu^-}{2}\right)_{(z=L)} = \mu_2 + \left(\frac{qIR_B}{2}\right).$$

These are exactly the new boundary conditions for the standard diffusion equation that we mentioned earlier (Eqs. (8.6)).

8.1.1 A disclaimer

The simple description provided above is an approximate one designed to convey a qualitative physical picture. The out-of-equilibrium occupation of different states is in general quite complicated and cannot necessarily be captured with just two potentials μ^+ and μ^-, *even for an elastic resistor at low bias*. Indeed the rest of this chapter is intended to give the reader a feeling for the underlying concepts and issues. In the next chapter we introduce the Boltzmann equation which is the gold standard for semiclassical transport against which all approximate pictures have to be measured. In subsequent chapters (Chapters 10, 11 and 12) we will discuss different aspects related to the difficult but very important concept of electrochemical potentials or *quasi-Fermi levels* under non-equilibrium conditions.

8.2 Electrochemical Potentials Out of Equilibrium

Related video lecture available at course website, Unit 3: L3.3.

As I mentioned earlier, it is conceptually straightforward to talk about electrochemical potentials inside the contacts which are large regions that always remain close to equilibrium and hence are described by Fermi functions (see Eq. (2.7)) with well-defined electrochemical potentials. But in an out-of-equilibrium channel, the electron distribution among the available states need not follow a Fermi function. In general one has to solve a full-fledged transport equation like the semiclassical Boltzmann equation to be introduced in the next chapter which allows us to calculate the full occupation factors $f(z; E)$. More generally for quantum transport one can use the non-equilibrium Green's function (NEGF) formalism discussed in Part B to solve for the quantum version of $f(z; E)$. Can we really represent these distribution functions using electrochemical potentials $\mu^+(z)$ and $\mu^-(z)$?

Interestingly for a perfectly ballistic channel with good contacts, such a representation in terms of $\mu^+(z)$ and $\mu^-(z)$ is exact and not just an approximation. All drainbound electrons (traveling along $+z$, see Fig. 8.4) are distributed according to the source contact with $\mu^+ = \mu_1$:

$$f^+(z; E) \;=\; f_1(E) \;\equiv\; \frac{1}{1 + \exp\left(\dfrac{E - \mu_1}{kT}\right)} \tag{8.15}$$

while all sourcebound electrons (traveling along $-z$) are distributed according to the drain contact with $\mu^- = \mu_2$:

$$f^-(z; E) \;=\; f_2(E) \;\equiv\; \frac{1}{1 + \exp\left(\dfrac{E - \mu_2}{kT}\right)}. \tag{8.16}$$

This is justified by noting that the drainbound channels from the source are filled only with electrons originating in the source and so these channels remain in equilibrium with the source with a distribution function $f_1(E)$. Similarly the sourcebound channels from the drain are in equilibrium with the drain with a distribution function $f_2(E)$.

Suppose at some energy $f_1(E) = 1$ and $f_2(E) = 0$ so that there are lots of electrons waiting to get out of the source, but none in the drain. We would then expect the drainbound lanes of the electronic highway to be completely full ("bumper-to-bumper traffic"), while the sourcebound lanes would all be empty as shown below in Fig. 8.4.

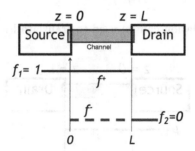

Fig. 8.4   Spatial profile of the occupation factors f^+ and f^- across a ballistic channel.

Of course this assumes that electrons do not turn around either along the way or at the ends. This means ballistic channels with good contacts where there are so many channels available that electrons can exit smoothly with a very low probability of turning around. If we either have bad contacts or diffusive channels, the solution in Eqs. (8.15) and (8.16) wouldn't work.

For diffusive channels with good contacts, Eqs. (8.15) and (8.16) suggest a plausible guess for what we might expect the distributions to look like in a diffusive channel. We assume the same Fermi-like function but with spatially varying electrochemical potentials reflecting the fact that electrons from the drainbound channels continually transfer over to the sourcebound lanes:

$$f^+(z; E) = \frac{1}{1 + \exp\left(\dfrac{E - \mu^+(z)}{kT}\right)} \tag{8.17a}$$

$$f^-(z; E) = \frac{1}{1 + \exp\left(\dfrac{E - \mu^-(z)}{kT}\right)}. \tag{8.17b}$$

If we accept these forms for the occupation factors, then it is straightforward to translate a plot of occupation factors f (like the one in Fig. 8.4) into a corresponding plot for the electrochemical potentials by noting that at low bias, the deviation of f from a reference value f_0 is proportional to the deviation of the corresponding μ from the corresponding reference value of μ_0:

$$f(E) - f_0(E) \approx \left(-\frac{\partial f_0}{\partial E}\right)(\mu - \mu_0) \quad \text{(same as Eq. (2.11))}.$$

For example, this relation can be used to translate Fig. 8.4 into Fig. 8.5.

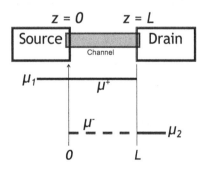

Fig. 8.5 Spatial profile of the electrochemical potentials μ^+ and μ^- across a ballistic channel, obtained from Fig. 8.4 by translating f's into μ's using Eq. (2.11).

Note that Eq. (8.17) is a "guess" that in general requires careful scrutiny and justification, if we are interested in quantitative results. For example, in an elastic resistor, every energy is independent and in general each one could exhibit a different spatial variation in the potential if the mean free path is energy-dependent. This means we should write the potentials as $\mu^{\pm}(z; E)$. We will talk further about similar issues in the next chapter when we discuss the Boltzmann equation.

8.3   Current from QFL's

Related video lecture available at course website, Unit 3: L3.4.

Let me finish up this chapter by establishing a key result that stated earlier without proof in Eq. (8.8). Usually we talk about the net current I which can be expressed as the difference between the drainbound flux I^{+} and the sourcebound flux I^{-}:

$$I(z) = I^{+}(z) - I^{-}(z). \tag{8.18}$$

The current I^{+} equals the amount of charge exiting from the right per unit time. In a time Δt, all the charge in a length $v_z \Delta t$ exits, so that

$$I^{+} = q \times (\textit{Electrons per unit length}) \times v_z.$$

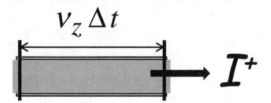

The number of electrons per unit length is equal to half the density of states (since only half the states carry current to the right) per unit length, $D(E)/2L$, times the fraction f^{+} of occupied states, so that

$$I^{+}(z; E) = q \underbrace{\frac{D(E)}{2L} \bar{u}(E)}_{M(E)/h} f^{+}(z; E).$$

Here $\bar{u}$ is the average v_z as defined in Eq. (4.8) and making use of the definition of the number of channels M from Eq. (4.13) we have

$$I^+(z;E) = q\frac{M(E)}{h}f^+(z;E). \tag{8.19}$$

Similarly

$$I^-(z;E) = q\frac{M(E)}{h}f^-(z;E). \tag{8.20}$$

This allows us to write the current from Eq. (8.18)

$$I(z) = \int_{-\infty}^{+\infty} dE(I^+(z;E) - I^-(z;E))$$

$$= \frac{q}{h}\int_{-\infty}^{+\infty} dE(f^+(z;E) - f^-(z;E))M(E). \tag{8.21}$$

Once again, to get from distribution functions $f^\pm$ to electrochemical potentials $\mu^\pm$, we make use of the low bias result (Eq. (2.11) to write

$$f^+(z;E) - f^-(z;E) = \left(-\frac{\partial f_0}{\partial E}\right)(\mu^+(z) - \mu^-(z)) \tag{8.22}$$

and obtain Eq. (8.8)

$$I(z) = \frac{q}{h}(\mu^+(z) - \mu^-(z)) \underbrace{\int_{-\infty}^{+\infty} dE\left(-\frac{\partial f_0}{\partial E}\right)M(E)}_{\equiv M} \tag{8.23}$$

provided we identify M with the thermally averaged $M(E)$ as indicated in Eq. (8.23).

Chapter 9

Boltzmann Equation

Related video lecture available at course website, Unit 3: L3.8.

9.1 Introduction

Interestingly so far in this book we have hardly ever mentioned the electric field, in contrast to most treatments of electronic transport which start by considering the electric field induced force as the driving term. It may seem paradoxical that we could obtain the conductivity without ever mentioning the electric field! Electric fields are typically visualized as the gradient of an electrostatic potential U/q. By contrast, we have been using the electrochemical potential μ as the basis for our discussions. It is important to recognize the difference between the two "potentials":

$$\underbrace{\mu}_{Electrochemical} = \underbrace{(\mu - U)}_{Chemical} + \underbrace{U}_{Electrostatic} \tag{9.1}$$

μ is a measure of the energy up to which the states are filled, while U determines the energy shift of the available states, so that $\mu - U$ is a measure of the degree to which the states are filled and hence the number of electrons. In the last chapter we obtained the equation

$$\frac{I}{A} = -\sigma_0 \frac{d(\mu/q)}{dz}. \tag{9.2}$$

But what we really showed was that

$$\frac{I}{A} = -\sigma_0 \frac{d(\mu - U)/q}{dz} \tag{9.3}$$

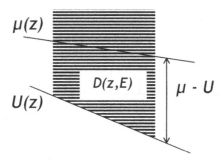

Fig. 9.1   The two potentials: Electrostatic U/q and electrochemical μ/q. $D(z;E)$ denotes the spatially varying density of states.

assuming zero electric field, $dU/dz = 0$. So how do we know what the correct equation is, when we include U?

It would seem that we needed to solve a whole new problem including the effect of the field $(= d(U/q)/dz)$ on electrons. However, this is unnecessary because the basic principles of equilibrium statistical mechanics require the current to be zero for a constant μ, just as there can be no heat current if the temperature is constant. Hence the current expression must have the form given in Eq. (9.2) which can be written as the sum of a drift term and a diffusion term

$$\frac{I}{A} = \underbrace{-\sigma_0 \frac{d(\mu - U)/q}{dz}}_{Diffusion} \underbrace{-\sigma_0 \frac{d(U/q)}{dz}}_{Drift} \qquad (9.4)$$

both of which must be described by the same coefficient σ_0, a requirement that leads to the Einstein relation between drift and diffusion. And that is why we can find σ_0 considering only the diffusion of electrons with $U = 0$, obtain Eq. (9.3) and just replace it with Eq. (9.2) which correctly accounts for "everything". There is really no need to work out the drift problem separately. What we called the diffusion equation is really the *drift-diffusion equation* even though we did not consider drift explicitly.

Couldn't we instead have neglected diffusion completely and just gone with the drift term? That way we could stick to the view that current is driven by electric fields and not have to bother with electrochemical potentials. The problem is that if we take this view then one has to invoke mysterious quantum mechanical forces to explain why all electrons are not affected by the field. In our discussion the energy window for transport (F_T, see Fig. 2.3) arises naturally from the difference in the "agenda" of the two contacts (see Eq. (2.10))

$$f_1(E) - f_2(E) = \left(-\frac{\partial f_0}{\partial E}\right)(\mu_1 - \mu_2)$$

as discussed in Chapters 2 and 3.

The point is that regardless of which potential we choose to work with, it finally affects transport through the occupation factor, f. In this chapter we will justify our neglect of drift more explicitly by introducing the Boltzmann Transport Equation (BTE) which is the standard starting point for all discussions of the transport of particles. We too could have used it as the starting point for but we did not do so because it is harder to digest with its multiple independent variables, compared to the ordinary differential equation in Chapter 8, which follows from relatively elementary arguments.

Even in this chapter we will not really do justice to the BTE. We will introduce it briefly and use it to show that for low bias, the current indeed depends only on $d\mu/dz$ and not on dU/dz, thus putting our discussion of steady-state, low bias transport without electric fields on a firmer footing and identifying possible issues with it.

Note the two qualifying phrases, namely "steady-state" and "low bias". We will show later in this chapter that for time varying transport, the neglect of electric fields can lead to errors, but we will not discuss it further in this book. However, even under steady-state conditions, electric fields can play an important role in determining the full current-voltage characteristics, once we go beyond low bias, as we saw in Chapter 7.

9.2 BTE from "Newton's Laws"

In Chapter 8 we introduced electron distribution functions $f^\pm$ and electrochemical potentials $\mu^\pm$ describing the drainbound and sourcebound currents $I^\pm$. Both the drainbound and sourcebound current, however, is composed of electrons traveling at different angles having different z-momemtum p_z, even though they all have the same energy (we are still talking about an elastic resistor) and hence the same total momentum. To include the effect of the electric field we need "momentum-resolved" distribution functions $f^\pm(z, p_z, t)$.

The BTE describes the evolution of such "momentum-resolved" distribution functions $f(z, p_z, t)$ that tell us the occupation of states with a given momentum p_z and velocity v_z at a location z at time t:

$$\frac{\partial f}{\partial t} + v_z\frac{\partial f}{\partial z} + F_z\frac{\partial f}{\partial p_z} = S_{op}f \tag{9.5}$$

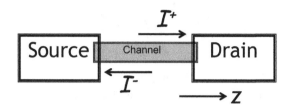

where F_z is the force on the electrons, and $S_{op}f$ symbolically represents the complex scattering processes that continually redistribute electrons among the available velocity states. The BTE with the right hand side set to zero (that is without scattering processes)

$$\frac{\partial f}{\partial t} + v_z \frac{\partial f}{\partial z} + F_z \frac{\partial f}{\partial p_z} = 0 \qquad (9.6)$$

is completely equivalent to describing a set of particles each with position $z(t)$ and momenta $p_z(t)$ that evolve according to the semiclassical laws of motion:

$$v_z \equiv \frac{dz}{dt} = \frac{\partial E}{\partial p_z} \qquad (9.7a)$$

$$F_z \equiv \frac{dp_z}{dt} = -\frac{\partial E}{\partial z} \qquad (9.7b)$$

where $E(z, p_z, t)$ is the total energy. Equations (9.7) describe semiclassical dynamics in single particle terms where the position $z(t)$ and momenta $p_z(t)$ for each of the electrons is a dependent variable evolving in time. By contrast, the BTE provides a collective description with all three independent variables z, p_z and t on an equal footing. To get from Eqs. (9.7) to (9.6) we start by noting that in the absence of scattering, we can write

$$f(z, \ p_z, \ t) = f(z - v_z \Delta t, \ p_z - F_z \Delta t, \ t - \Delta t)$$

reflecting the fact that any electron with a momentum p_z at z at time t, must have had a momentum of $p_z - F_z \Delta t$, at $z - v_z \Delta t$ a little earlier at time $t - \Delta t$.

Next we expand the right hand side to the first term in a Taylor series to write

$$f(z, p_z, t) = f(z, p_z, t) - \frac{\partial f}{\partial z} v_z \Delta t - \frac{\partial f}{\partial p_z} F_z \Delta t - \frac{\partial f}{\partial t} \Delta t$$

Eq. (9.6) follows readily on canceling out the common terms. The left hand side of the BTE thus represents an alternative way of expressing the laws of motion. What makes it different from mere mechanics, however, is the stochastic scattering term on the right which makes the distribution function f approach the equilibrium Fermi function when external driving terms are absent. This last point of course is not meant to be obvious. It requires an extended discussion of the scattering operator S_{op} that we talk a little more about in Chapter 15 when we discuss the second law. For our purpose it suffices to note that a common approximation for the scattering term is the *relaxation time approximation* (RTA)

$$S_{op}f \cong -\frac{f-\overline{f}}{\tau} \qquad (9.8)$$

which assumes that the effect of the scattering processes is proportional to the degree to which a given distribution f differs from the *local equilibrium distribution* $\overline{f}$.

One comment about why we call this approach *semiclassical*. The BTE is *classical* in the sense that it is based on a particle view of electrons. But it is not *fully* classical, since it typically includes quantum input both in the scattering operator S_{op} and in the form of the energy-momentum relation. For example, graphene is often described by a linear energy-momentum relation

$$\mathbf{E} = \nu_0\mathbf{p}$$

a result that is usually justified in terms of the bandstructure of the graphene lattice requiring quantum mechanics that came after Boltzmann's time. This is also why we put Newton's laws within quotes in the section title. But once we accept this extension, many transport properties of graphene can be understood in classical particulate terms using the BTE that Boltzmann taught us to use.

9.3 Diffusion Equation from BTE

We start by combining the RTA (Eq. (9.8)) with the full BTE (Eq. (9.5)) to obtain for steady-state ($\partial/\partial t = 0$),

$$v_z\frac{\partial f}{\partial z} + F_z\frac{\partial f}{\partial p_z} = -\frac{f-\overline{f}}{\tau}. \qquad (9.9)$$

In the presence of an electric field we can write the total energy as

$$E(z, p_z) = \varepsilon(p_z) + U(z) \qquad (9.10)$$

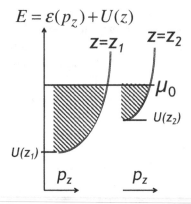

Fig. 9.2 The energy momentum relation with $U = 0$ is shifted locally by $U(z)$. At equilibrium the electrochemical potential μ_0 is spatially constant.

where $\varepsilon(p_z)$ denotes the energy-momentum relation with $U = 0$ and this gets shifted locally by $U(z)$ as sketched in Fig. 9.2.

The first point to note is that the equilibrium distribution with a constant electrochemical potential μ_0

$$f_0(z, p_z) \; = \; \cfrac{1}{\exp\left(\cfrac{E(z, p_z) - \mu_0}{kT}\right) + 1} \tag{9.11}$$

satisfies the BTE in Eq. (9.9). The right hand side of Eq. (9.9) is zero simply because $\overline{f} = f_0$, but it takes a little differential calculus to see that the left hand side is zero too. Defining

$$X_0 \; \equiv \; E(z, p_z) \; - \; \mu_0 \; = \; \varepsilon(p_z) + U(z) - \mu_0 \tag{9.12}$$

we have

$$v_z \frac{\partial f_0}{\partial z} \; + \; F_z \frac{\partial f_0}{\partial p_z} = \left(\frac{\partial f_0}{\partial X_0} \right) \left(v_z \frac{\partial X_0}{\partial z} \; + \; F_z \frac{\partial X_0}{\partial p_z} \right)$$

$$= \left(\frac{\partial f_0}{\partial X_0} \right) \left(v_z \frac{\partial E}{\partial z} \; + \; F_z \frac{\partial E}{\partial p_z} \right) = 0$$

making use of Eqs. (9.7).

Out of equilibrium, we assume two separate distribution functions $f^\pm(z, p_z)$ as in Eqs. (8.17a) and (8.17b) for the right-moving ($v_z > 0$) and the left-moving ($v_z < 0$) electrons with separate spatially varying electrochemical potentials $\mu^\pm(z)$:

$$f^\pm(z, p_z) \; = \; \cfrac{1}{\exp\left(\cfrac{E(z, p_z) - \mu^\pm(z)}{kT}\right) + 1} \tag{9.13}$$

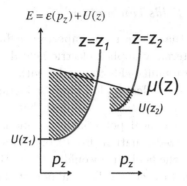

Fig. 9.3 Same as Fig. 9.2, but the electrochemical potential $\mu(z)$ varies spatially reflecting a non-equilibrium state.

Using Eq. (9.13), the left hand side of BTE (see Eq. (9.9)) reduces to

$$\left(\frac{\partial f}{\partial X}\right)\left(v_z\frac{\partial X}{\partial z} + F_z\frac{\partial X}{\partial p_z}\right) = \left(\frac{\partial f}{\partial X}\right)\left(-v_z\frac{\partial \mu^\pm}{\partial z}\right) \tag{9.14}$$

where

$$X^\pm \equiv E(z, p_z) - \mu^\pm(z).$$

We now assume small deviations in $\mu^\pm(z)$ from the *local equilibrium value* so that we can write the left hand side as

$$\left(\frac{\partial f}{\partial X}\right)_{X=\overline{X}}\left(-v_z\frac{\partial \mu^\pm}{\partial z}\right)$$

and use our standard Taylor series expansion (see Eq. (2.11)) to write the right hand side of BTE as

$$-\frac{f^\pm - \overline{f}}{\tau} \approx \left(\frac{\partial f}{\partial X}\right)_{X=\overline{X}}\frac{\mu^\pm - \overline{\mu}}{\tau}.$$

Combining the two sides

$$v_z\frac{\partial \mu^\pm}{\partial z} = -\frac{\mu^\pm - \overline{\mu}}{\tau}. \tag{9.15}$$

This gives us two separate equations for the two electrochemical potentials μ^+ and μ^- for the right-moving ($v_z > 0$) and left-moving ($v_z < 0$) electrons

$$\frac{\partial \mu^+}{\partial z} = -\frac{\mu^+ - \overline{\mu}}{|v_z|\tau}, \quad \frac{\partial \mu^-}{\partial z} = -\frac{\mu^- - \overline{\mu}}{-|v_z|\tau}.$$

Assuming $\overline{\mu} = (\mu^+ + \mu^-)/2$, we obtain

$$\frac{\partial \mu^+}{\partial z} = -\frac{\mu^+ - \mu^-}{\lambda} = \frac{\partial \mu^-}{\partial z} \tag{9.16}$$

with $\lambda = 2|v_z|\tau$. Combining with Eq. (8.10) for the current, we obtain the result (Eq. (8.13)) stated without proof in Chapter 8. Note that we have included electric fields explicitly and shown that their effect cancels out.

9.3.1 *Equilibrium fields can matter*

We believe, however, that there is an important subtlety worth pointing out. Although the externally applied electric field does not affect the low bias conductance, any inbuilt fields that exist within the conductor under equilibrium conditions can affect its low bias conductance. Let me explain. Note that in our treatment above we assumed that under non-equilibrium conditions, the electrochemical potential is a function of z (Eq. (9.13)) and the resulting linearized equation (Eq. (9.16)) does not involve the field $F_z = dU/dz$. However, the field term would not have dropped out so nicely if we were to assume that the electrochemical potential is not just a function of z, but of both z and p_z. Instead of Eq. (9.15) we would then obtain

$$v_z \frac{\partial \mu}{\partial z} + F_z \frac{\partial \mu}{\partial p_z} = -\frac{\mu - \overline{\mu}}{\tau}. \tag{9.17}$$

However, the additional term involving the field F_z does not play a role in determining linear conductivity because it is $\sim V^2$, V being the applied voltage. At equilibrium with $V = 0$, $\mu = \overline{\mu}$, so that both derivatives appearing on the left are zero. Under bias, in principle, both are non-zero. But the point is that while v_z is a constant, the applied field F_z is also $\sim V$. So while the first term on the left is $\sim V$, the second term is $\sim V^2$.

But this argument would not hold if F_z were not the applied field, but internal inbuilt fields independent of V that are present even at equilibrium. Equilibrium requires a constant μ and NOT a constant U. The equilibrium condition depicted in Fig. 9.2 is quite common in real conductors, with varying $U(z)$ corresponding to non-zero fields F_z. Indeed this picture could also represent an interface between dissimilar materials (called "heterostructures") where the discontinuity in band edges is often modeled with effective fields.

The point is that such equilibrium fields can affect the low bias conductance. For an ideal homogeneous conductor we do not have such fields. But even then we need to make two contacts in order to measure the resistance. Each such contact represents a heterostructure qualitatively similar to that shown in Fig. 9.2 with inbuilt effective (if not real) fields. In Chapter 11 we will discuss the Hall effect where we have to keep the F_z term in Eq. (9.17) in order to account for the presence of an external magnetic field.

9.4 The Two Potentials

In this book we will generally focus on steady-state transport involving the injection of electrons from a source and their collection by a drain (Fig. 9.4).

We have seen that the low bias conductance can be understood in terms of the electrochemical potential μ, without worrying about the electrostatic potential U. However, we would like to briefly consider ac transport through a nanowire far from any contacts where we have a local voltage $V(z,t)$ and current $I(z,t)$ (Fig. 9.5), because this provides a contrasting example where it is important to pay attention to the difference between the two potentials even for low bias, in order to obtain the correct inductance and capacitance.

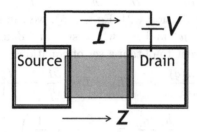

Fig. 9.4   So far we have talked of steady-state transport involving the injection of electrons by a source and their collection by a drain contact.

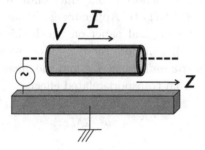

Fig. 9.5   AC or time varying transport along a nanowire can be described in terms of a voltage $V(z,t)$ and a current $I(z,t)$.

For this problem too we start from the BTE with the RTA approximation as in the last section, but we do not set $\partial/\partial t = 0$. Instead we start from

$$\frac{\partial f}{\partial t} + v_z\frac{\partial f}{\partial z} + F_z\frac{\partial f}{\partial p_z} = -\frac{f - \overline{f}}{\tau}$$

and linearize it assuming a distribution of the form (compare Eq. (9.13))

$$f(z, p_z, t) = \frac{1}{\exp\left(\dfrac{E(z, p_z, t) - \mu(z, t)}{kT}\right) + 1}. \tag{9.18}$$

Compared to the steady-state problem (Eq. (9.15)) we now have two extra terms involving the time derivatives of E and μ:

$$\frac{\partial \mu}{\partial t} + v_z \frac{\partial \mu}{\partial z} - \frac{\partial E}{\partial t} = -\frac{\mu - \bar{\mu}}{\tau}. \tag{9.19}$$

As we did in the last Section with Eq. (9.15), we can separate Eq. (9.19) into two equations for μ^+ and μ^-, whose sum and difference are identified with voltage and current to obtain a set of equations

$$\frac{\partial(\mu/q)}{\partial z} = -(L_K + L_M)\frac{\partial I}{\partial t} - \frac{I}{\sigma A} \tag{9.20a}$$

$$\frac{\partial(\mu/q)}{\partial t} = -\left(\frac{1}{C_Q} + \frac{1}{C_E}\right)\frac{\partial I}{\partial z} \tag{9.20b}$$

that look just like the transmission line equations with a distributed series inductance and resistance and a shunt capacitance.

The algebra getting from Eq. (9.19) to Eqs. (9.20) is a little long-winded and since time-varying transport is only incidental to our main message we have relegated the details to Appendix E. Those who are really interested can look at the original paper on which this discussion is based (Salahuddin *et al.*, 2005). But note the two inductors and the two capacitors in series. The *kinetic inductance* L_K and the *quantum capacitance* C_Q per unit length, arise from transport-related effects

$$L_K = \frac{h}{q^2} \frac{1}{\langle 2Mv_z \rangle} \tag{9.21a}$$

$$C_Q = \frac{q^2}{h} \left\langle \frac{2M}{v_z} \right\rangle \tag{9.21b}$$

while the L_M and the C_E are just the normal *magnetic inductance* and the *electrostatic capacitance* from the equations of magnetostatics and electrostatics. The point I wish to make is that the fields enter the expression for the energy $E(z, p_z, t)$ and if we ignore the fields we would miss the $\partial E/\partial t$ term in Eq. (9.19) to obtain

$$\frac{\partial \mu}{\partial t} + v_z \frac{\partial \mu}{\partial z} = -\frac{\mu - \bar{\mu}}{\tau} \tag{9.22}$$

and after working through the algebra obtain instead of Eqs. (9.20)

$$\frac{\partial(\mu/q)}{\partial z} = -L_K \frac{\partial I}{\partial t} - \frac{I}{\sigma A} \tag{9.23a}$$

$$\frac{\partial(\mu/q)}{\partial t} = -\frac{1}{C_Q} \frac{\partial I}{\partial z}. \tag{9.23b}$$

Do these equations approximately capture the physics? Not unless we are considering wires with very small cross-sections so that M is a small number making $L_K \gg L_M$ and $C_Q \ll C_E$. We could recover the correct answer from Eqs. (9.23) by replacing the μ with $\mu - U$ and then using the laws of electromagnetics to replace

$$\frac{\partial U}{\partial t} \text{ with } \frac{1}{C_E} \frac{\partial I}{\partial z} \quad \text{and} \quad \frac{\partial U}{\partial z} \text{ with } L_M \frac{\partial I}{\partial t}.$$

But these replacements may not be obvious and it is more straightforward to go from Eq. (9.19) to (9.20) as spelt out in Appendix E. Note that if we specialize to steady-state ($\partial/\partial t = 0$), both Eqs. (9.19) and (9.22) give us back our old diffusion equation (Eq. (8.2)). As we argued earlier, for low bias steady-state transport, the applied electric field can be treated as incidental.

Chapter 10

Quasi-Fermi Levels

10.1 Introduction

Electrochemical potentials have played an important role in our discussion, starting from Chapter 2 where I stressed that electron flow is driven by the difference in the electrochemical potentials μ_1 and μ_2 in the two contacts. However, talking about electrochemical potentials inside the channel, as we did later in Chapter 8 when discussing the diffusion equation, often raises eyebrows. This is because an electrochemical potential of μ implies that the occupation of all available states are described by the corresponding Fermi function (Eq. (2.2))

$$f(E) = \frac{1}{1 + \exp\left(\dfrac{E - \mu}{kT}\right)}.$$

This is approximately true of large contacts which always remain close to equilibrium, but not necessarily true of small conductors even for small applied voltages. As we saw in Chapter 8, it was important to introduce two separate electrochemical potentials μ^+ and μ^- in order to understand the interface resistance that is the key feature of the new Ohm's law (Eq. (4.5)).

Non-equilibrium electrochemical potentials of this type can be very useful in understanding current flow and is widely used by device engineers. It is common to use two different potentials (often called quasi-Fermi levels) for conduction and valence bands and in Chapter 12 we will talk about other examples of quasi-Fermi levels and argue that controlling such potentials with creatively designed "smart" contacts could lead to unique devices.

In spite of the obvious utility of the concept, many experts are uneasy about invoking non-equilibrium electrochemical potentials inside nanoscale devices which they view as ill-defined concepts that cannot be measured.

Instead they feel conceptually on solid ground by sticking to terminal descriptions in terms of the electrochemical potentials at the contacts. In this chapter, I would like to address some of these issues related to non-equilibrium potentials and their measurability using a simple example which will also allow us to connect our discussion to the Landauer formulas and the Büttiker formula that form the centerpiece of the transmission formalism widely used in mesoscopic physics.

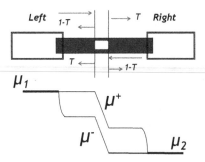

Fig. 10.1   Potential variation across a defect.

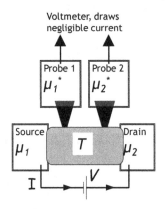

Fig. 10.2   Four-terminal measurement of conductance of an otherwise ballistic one-dimensional conductor having a single "defect" in the middle, through which electrons have a probability T of transmitting.

So far we have talked about normal resistors with uniformly distributed scatterers characterized by a mean free path. Instead, following Landauer, let us consider an otherwise ballistic channel with a single localized defect that lets a fraction T of all the incident electrons proceed along the original direction, while the rest $1 - T$ get turned around (see Fig. 10.1). We could follow our arguments from Chapter 8 to obtain the spatial variation of the potentials μ^+ and μ^- across the scatterer, and use it to deduce the resistance of the scatterer. But experts are often uneasy about non-equilibrium potentials and one way to bypass these questions is to consider a four-terminal measurement (Fig. 10.2) using two additional voltage probes that draw negligible current, to measure the voltage drop across the defect. We will show that if the voltage probes are identical and weakly coupled (non-invasive) then this four-terminal conductance G_{4t} is given by

$$G_{4t} = \frac{I}{(\mu_1^* - \mu_2^*)/q} = M \frac{q^2}{h} \frac{T}{1 - T} \qquad (10.1)$$

M being the number of channels or modes in the conductor introduced at the end of Chapter 4. But if we were to determine the conductance using the actual voltage applied to the current-carrying terminals we would obtain a lower conductance:

$$G_{2t} = \frac{I}{(\mu_1 - \mu_2)/q} = M \frac{q^2}{h} T. \qquad (10.2)$$

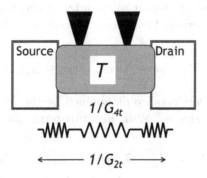

Fig. 10.3 The two-terminal resistance can be viewed as the four-terminal resistance in series with the interface resistance.

The difference between the two-terminal (Eq. (10.2)) and four-terminal (Eq. (10.1)) resistances reflects the same *interface resistance*

$$\frac{1}{G_{2t}} - \frac{1}{G_{4t}} = \frac{h}{q^2 M}$$

that differentiates the new Ohm's law (Eq. (4.5)) from the standard one (Eq. (1.1)).

Although the interface resistance was recognized for metallic resistors in the late 1960s and is known as the *Sharvin resistance*, its ubiquitous role is not widely appreciated even today. In the early 1980s there was considerable confusion and discussion about the difference between the two conductance formulas in Eqs. (10.1) and (10.2) and *Imry* is credited with identifying the difference as a quantized Sharvin resistance related to the interfaces. With the rise of mesoscopic physics, Eq. (10.2) has come to be widely used and known as the Landauer formula while Landauer's original formula (Eq. (10.1)) is relatively forgotten, and not many recognize the difference.

The reader may wonder why the four-terminal Landauer formula came to be "forgotten". After all resistance measurements are commonly made in the four terminal configuration in order to exclude any contact resistance. Don't such measurements require Eq. (10.1) for their interpretation? Sort of, but not exactly. Let me explain.

One problem in the early days of mesoscopic physics was that the voltage probes were strongly coupled to the main conductor and behaved like "additional defects" whose effect could not simply be ignored. In order to interpret real experiments using four-terminal configurations, Büttiker (see Büttiker 1988) found an elegant solution by writing the current I_m at terminal m of a multi-terminal conductor in terms of the terminal potentials μ_n:

$$I_m = \frac{1}{q} \sum_m G_{m,n} \left(\mu_m - \mu_n \right) \tag{10.3}$$

where $G_{m,n}$ is the conductance determined by the transmission $T_{m,n}$ between terminals m and n. With just two terminals, Büttiker's formula reduces to

$$I_1 = (1/q)\, G_{12} \left(\mu_1 - \mu_2 \right) = -I_2$$

which is the same as the two-terminal Landauer formula (Eq. (10.2)) if we identify G_{12} as $(q^2/h)M$. But the power of Eq. (10.3) lies in its ability to provide a quantitative basis for the analysis of multi-terminal structures like the one in Fig. 10.2.

Knowing $G_{m,n}$, if we knew all the potentials μ_m, we could use Eq. (10.3) to calculate the currents I_m at all the terminals. Of course for the voltage probes 1* and 2* we do not know the voltages they will float to and so we do not know μ_1^* or μ_2^*, to start with. But we do know the currents I_1^* and I_2^*, each of which must be zero, since the high impedance voltmeter draws negligible current.

The point is that if we know either μ_m or I_m at each terminal m we can solve Eq. (10.3) to obtain whatever we do not know. In this chapter we will look at a specific problem, namely the voltage drop across a defect (Fig. 10.1) and show that with weakly coupled non-invasive probes the Büttiker formula indeed gives the same answers as we get by looking directly at the electrochemical potentials μ^+ and μ^- inside the conductor. This is reassuring because the approach due to Büttiker deals directly with measurable terminal quantities and so appears conceptually on more comfortable ground.

The development of scanning probe microscopy (SPM) has made it possible to use nanoscale tunneling contacts as voltage probes whose effect is indeed negligible. Measurements using such "non-invasive" probes do provide experimental support for the four-terminal Landauer formula, but there is a subtlety involved.

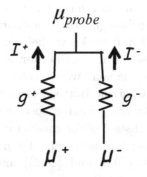

Fig. 10.4   Simple circuit model for a voltage probe.

What a voltage probe measures is some weighted average of the two potentials μ^+ and μ^-. The exact weighting depends on the construction of the probes. We could model it by associating conductances g^+ and g^- with the transmission of electrons from the + and the − streams into the probes respectively. Setting the net probe current to zero we can write

$$g^+ \left(\mu^+ - \mu_{probe}\right) + g^- \left(\mu^- - \mu_{probe}\right) = 0$$

so that

$$\mu_{probe} = \underbrace{\frac{g^+}{g^+ + g^-}}_{\alpha} \mu^+ + \underbrace{\frac{g^-}{g^+ + g^-}}_{1-\alpha} \mu^-. \tag{10.4}$$

For atomic scale probes that are much smaller than an electron wavelength we expect the two conductances to be similar so that the weighting factor $\alpha \sim 50\%$, so that the probe measures the average potential

$$\mu_{probe} = (\mu^+ + \mu^-)/2.$$

For larger probes, however, it is possible for a voltage probe to have a pronounced bias for one stream or the other leading to a weighting factor α different from 50%. If this weighting happens to be different for the two probes 1* and 2*, it could change the measured resistance from that predicted by Eq. (10.1). Indeed, experimental measurements have even shown *negative resistance*, something that cannot be understood in terms of Eq. (10.1).

However, some of this is due to quantum interference effects that make the simple semiclassical description in terms of $\mu^{\pm}$ inadequate as we will see in Part B. However, one could use a more sophisticated quantum version of Eq. (10.4) or use the Büttiker formula, with the conductances $G_{m,n}$ calculated from an appropriate quantum transport model.

The bottom line is that if we know the correct internal state of the conductor in terms of a set of non-equilibrium electrochemical potentials, we can predict what a specific non-invasive voltage probe will measure and the result should match what the Büttiker formula predicts. The reverse, however, is not true. Knowing what a specific probe will measure, we cannot deduce the internal state of the conductor.

With that rather long "introduction" let us now look at the two Landauer formulas (Eqs. (10.1) and (10.2)) and the Büttiker formula (Eq. (10.3)) in a little more detail.

10.2 The Landauer Formulas (Eqs. (10.1) and (10.2))

Related video lecture available at course website, Unit 3: L3.5.

Getting back to the problem of finding the potential variation across a defect in an otherwise ballistic conductor (Fig. 10.1), we start by relating the outgoing currents to the incoming currents as follows

$$I^+(Right) = TI^+(Left) + (1-T)I^-(Right)$$

$$I^-(Left) = (1-T)I^+(Left) + TI^-(Right).$$

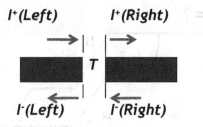

We can then convert the currents to occupation factors (see Eqs. (8.19) and (8.20))

$$f^+(Right) = Tf^+(Left) + (1-T)f^-(Right)$$

$$f^-(Left) = (1-T)f^+(Left) + Tf^-(Right)$$

and then to potentials using the Taylor's series (Eq. (2.10)) argument as before

$$\mu^+(Right) = T\mu^+(Left) + (1-T)\mu^-(Right)$$
$$= T\mu_1 + (1-T)\mu_2 \qquad (10.5)$$

$$\mu^-(Left) = (1-T)\mu^+(Left) + T\mu^-(Right)$$
$$= (1-T)\mu_1 + T\mu_2. \qquad (10.6)$$

The algebra can be simplified by choosing the potential for one of the contacts as zero and the other as one. The actual potential differences can then be obtained by multiplying by the actual $\mu_1 - \mu_2 = qV$.

Equations (10.5) and (10.6) then give us the picture shown in Fig. 10.5 leading to

$$\mu^+ - \mu^- = T(\mu_1 - \mu_2)$$

as long as both μ^+ and μ^- are evaluated at the same location on the left or on the right of the scatterer.

Using

$$I = \frac{q}{h}M(\mu^+ - \mu^-) \quad \text{(see Eq. (8.8))}$$

for the current we obtain the standard Landauer formula (Eq. (10.2)). To obtain the first Landauer formula we find the drop in either μ^+ or μ^- across the scatterer:

$$\mu^+(Left) - \mu^+(Right) = (1-T)(\mu_1 - \mu_2) \qquad (10.7a)$$

$$\mu^-(Left) - \mu^-(Right) = (1-T)(\mu_1 - \mu_2) \qquad (10.7b)$$

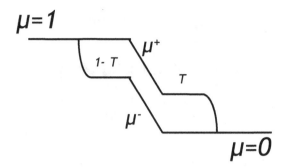

Fig. 10.5 Spatial profile of μ^+ and μ^- across a scatterer normalized to an overall potential difference of one. The actual potential differences can be obtained by multiplying by the actual $\mu_1 - \mu_2 = qV$.

and then divide the current by it to obtain the result stated in Eq. (10.1):

$$G_{4t} = \frac{q^2}{h} M \frac{T}{1-T}.$$

Note, however, that we are looking at the electrochemical potentials inside the conductor. How does this relate to the voltage measured by non-invasive voltage probes implemented using scanning tunneling probes (STP)?

Assuming that the probe measures the average of μ^+ and μ^- we obtain the plot shown in Fig. 10.6 using the results from Fig. 10.5. What if the probe measures a weighted average of μ^+ and μ^- with some α (see Eq. (10.4)) other than 50%? As long as α is the same for both probes, the drop across the scatterer would still be given by

$$\mu_{probe}(Left) - \mu_{probe}(Right) = (1-T)(\mu_1 - \mu_2) \qquad (10.8)$$

thus leading to the same Landauer formula (Eq. (10.1)). But if the weighting factor α were different for the two probes then the result would not match Eq. (10.1). As an extreme example if α were zero on the left and one on the right,

$$\mu_{probe}(Left) - \mu_{probe}(Right) = (1-2T)(\mu_1 - \mu_2)$$

leading to a negative resistance for $T > 0.5$.

Clearly the concept of non-equilibrium potentials μ^+ and μ^- should be used wisely with caution. But it does lead to intuitively understandable results. The potential drops across the defect but not across the ballistic regions, suggesting that the defect represents a resistance given by Eq. (10.1). Note, however, that we are still talking about elastic resistors. We have an IR drop in the voltage, but no corresponding I^2R in Joule dissipation. All dissipation is still in the contacts.

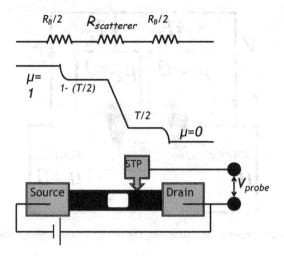

Fig. 10.6   A scanning tunneling probe (STP) measures the average electrochemical potential $(\mu^+ + \mu^-)/2$.

10.3   Büttiker Formula (Eq. (10.3))

Related video lecture available at course website, Unit 3: L3.6.

Equation (10.3) deals directly with the experimentally measured terminal quantities bypassing any questions regarding the internal variables. The point we wish to stress is the general applicability of this result irrespective of whether the resistor is elastic or not. Indeed, as we will see we can obtain it invoking very little beyond linear circuit theory. We start by defining a multi-terminal conductance

$$G_{m,n} \equiv -\frac{\partial I_m}{\partial(\mu_n/q)}, \ m \neq n \tag{10.9a}$$

$$G_{m,m} \equiv +\frac{\partial I_m}{\partial(\mu_m/q)}. \tag{10.9b}$$

Why do we have a negative sign for $m \neq n$, but not for $m = n$? The motivation can be appreciated by looking at a representative multi-terminal structure (Fig. 10.7). An increase in μ_1 leads to an incoming or positive current at terminal 1, but leads to *outgoing* or negative currents at the other terminals. The signs in Eqs. (10.9) are included to make the coefficients come out positive as we intuitively expect a conductance to be.

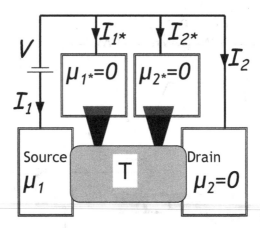

Fig. 10.7 Thought experiment based on the four-terminal measurement set-up in Fig. 10.2.

In terms of these conductance coefficients, we can write the current as

$$I_m = G_{m,m}\frac{\mu_m}{q} - \sum_{n\neq m} G_{m,n}\frac{\mu_n}{q}. \qquad (10.10)$$

The conductance coefficients must obey two important "sum rules" in order to meet two important conditions. Firstly, the currents predicted by Eq. (10.10) must all be zero if all the μ's are equal, since there should be no external currents at equilibrium. This requires that

$$G_{m,m} = \sum_{n\neq m} G_{m,n}. \qquad (10.11)$$

Secondly, for any choice of μ's, the currents I_m must add up to zero. This requires that

$$G_{m,m} = \sum_{n\neq m} G_{n,m} \qquad (10.12)$$

but it takes a little algebra to see this from Eq. (10.10). First we sum over all m

$$\sum_m I_m = 0 = \sum_m G_{m,m}\frac{\mu_m}{q} - \sum_m \sum_{n\neq m} G_{m,n}\frac{\mu_n}{q}$$

and interchange the indices n and m for the second term to write

$$0 = \sum_m G_{m,m}\frac{\mu_m}{q} - \sum_m \sum_{n\neq m} G_{n,m}\frac{\mu_m}{q}$$

which can be true for all choices of μ_m only if Eq. (10.12) is satisfied. We can combine Eqs. (10.11) and (10.12) to obtain the "sum rule" succinctly:

$$G_{m,m} = \sum_{n \neq m} G_{m,n} = \sum_{n \neq m} G_{n,m}. \tag{10.13}$$

Making use of the sum rule (Eq. (10.13)) we can re-write the first term in Eq. (10.10) to obtain Eq. (10.3):

$$I_m = \left(\frac{1}{q}\right) \sum_n G_{m,n}(\mu_m - \mu_n) \text{ (same as Eq. (10.3))}.$$

Note that it is not necessary to restrict the summation to $n \neq m$, since the term with $n = m$ is zero anyway. An alternate form that is sometimes useful is to write

$$I_m = \sum_n g_{m,n} \frac{\mu_n}{q} \tag{10.14}$$

where the response coefficients defined as

$$g_{m,n} \equiv -G_{m,n}, \; m \neq n \tag{10.15}$$

$$g_{m,m} \equiv G_{m,m}. \tag{10.16}$$

The sum rule in Eq. (10.13) can be rewritten in term of this new response coefficient:

$$\sum_n g_{m,n} = \sum_n g_{n,m} = 0. \tag{10.17}$$

10.3.1   *Application*

In Section 10.2 we analyzed the potential profile across a single scatterer with transmission probability T. Based on this discussion we would expect that two non-invasive probes inserted before and after the scatterer should float to potentials $1 - (T/2)$ and $T/2$ as indicated in Fig. 10.8. But will Büttiker's approach get us the same result?

We start from Eq. (10.14) noting that we have four currents and four potentials, labeled 1, 2, 1* and 2*:

$$\begin{Bmatrix} I_1 \\ I_2 \\ I_1^* \\ I_2^* \end{Bmatrix} = \frac{Mq}{h} \begin{bmatrix} \mathbf{A} \; \mathbf{B} \\ \mathbf{C} \; \mathbf{D} \end{bmatrix} \begin{Bmatrix} \mu_1 \\ \mu_2 \\ \mu_1^* \\ \mu_2^* \end{Bmatrix} \tag{10.18}$$

where $\mathbf{A}$, $\mathbf{B}$, $\mathbf{C}$ and $\mathbf{D}$ are each (2×2) matrices.

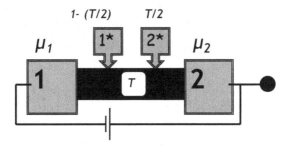

Fig. 10.8 Based on Fig. 10.6, we expect that two non-invasive probes inserted before and after a scatterer with transmission probability T to float to potentials $1 - (T/2)$ and $T/2$ respectively.

$$\left\{ \begin{matrix} I_1^* \\ I_2^* \end{matrix} \right\} = \left\{ \begin{matrix} 0 \\ 0 \end{matrix} \right\}.$$

Since we have

$$\left\{ \begin{matrix} \mu_1^* \\ \mu_2^* \end{matrix} \right\} = -\mathbf{D}^{-1}\mathbf{C} \left\{ \begin{matrix} \mu_1 \\ \mu_2 \end{matrix} \right\}. \tag{10.19}$$

Now we can write $\mathbf{C}$ and $\mathbf{D}$ in the augmented form

$$[\mathbf{C} \ \mathbf{D}] = \begin{bmatrix} -t_1 & -t_2 & r & 0 \\ -t_2' & -t_1' & 0 & r' \end{bmatrix} \tag{10.20}$$

where the elements t_1, t_2, t_1' and t_2' of the matrix $[\mathbf{C} \ \mathbf{D}]$ can be visualized as the probability that an electron transmits from 1 to 1*, 2 to 1*, 2 to 2* and 1 to 2* respectively as sketched in Fig. 10.9. We have assumed that both probes 1* and 2* are weakly coupled so that any direct transmission between them can be ignored.

The sum rule in Eq. (10.17) then requires that

$$r = t_1 + t_2 \tag{10.21a}$$

$$r' = t_2' + t_1'. \tag{10.21b}$$

This yields

$$\mu_1^* = \frac{t_1}{t_1 + t_2}\mu_1 + \frac{t_2}{t_1 + t_2}\mu_2 \tag{10.22a}$$

$$\mu_2^* = \frac{t_2'}{t_1' + t_2'}\mu_1 + \frac{t_1'}{t_1' + t_2'}\mu_2. \tag{10.22b}$$

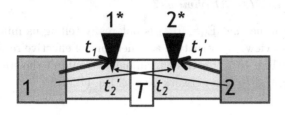

Fig. 10.9 The elements t_1, t_2, t_1' and t_2' of the matrix $[\mathbf{C}\ \mathbf{D}]$ can be visualized as the probability that an electron transmits from 1 to 1*, 2 to 1*, 2 to 2* and 1 to 2* respectively.

So far we have kept things general, making no assumptions other than that of weakly coupled probes. Now we note that for our problem (Fig. 10.9), t_1 can be written as

$$t_1 = \tau + (1 - T)\tau \qquad (10.23)$$

since an electron from 1 has a probability of τ to get into probe 1* directly plus a probability of $1 - T$ times τ to get reflected from the scatterer and then get into probe 1*. Similarly t_2 can be written as

$$t_2 = T\tau \qquad (10.24)$$

since an electron from 2 has to cross the scatterer (probability T) and then enter the weakly coupled probe 1* (probability τ). Similarly we can argue $t_1 = t_1'$, $t_2 = t_2'$. Using these results in Eqs. (10.22) and setting $\mu_1 = 1$, $\mu_2 = 0$, we then obtain,

$$\mu_1^* = 1 - (T/2) \qquad (10.25a)$$

$$\mu_2^* = T/2 \qquad (10.25b)$$

in agreement with what we expected from the last section (Fig. 10.8). As mentioned earlier, this is reassuring since the Büttiker formula deals only with terminal quantities, bypassing the subtleties of non-equilibrium electrochemical potentials.

However, the real strength of Eq. (10.3) lies in its model-independent generality. It should be valid in the linear response regime for all conductors, simple and complex, large and small. The conductances $G_{m,n}$ in Eq. (10.3) can be calculated from a microscopic transport model like the Boltzmann equation introduced in Chapter 9 or the quantum transport model discussed in Part B. Sometimes they can even be guessed and as long as we are careful about not violating the sum rules we should get reasonable results.

10.3.2 *Is Eq. (10.3) obvious?*

Some might argue that Eq. (10.3) is not really telling us much. After all, we can always view any structure as a network of effective resistors like the one shown in Fig. 10.10 for three terminals? Wouldn't the standard circuit equations applied to this network give us Eq. (10.3)?

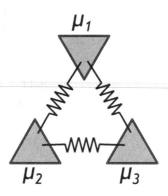

Fig. 10.10 The Büttiker formula (Eq. (10.3)) can be visualized as a network of resistors, only if the conductances are reciprocal, that is, if $G_{m,n} = G_{n,m}$.

The answer is "yes" if we consider only normal circuits for which electrons transmit just as easily from m to n as from n to m so that

$$G_{m \leftarrow n} = G_{n \leftarrow m}$$

where we have added the arrows in the subscripts to denote the standard convention for the direction of electron transfer. Equation (10.3), however, goes far beyond such normal circuits and was intended to handle conductors in the presence of magnetic fields for which

$$G_{m \leftarrow n} \neq G_{n \leftarrow m}.$$

For such conductors, Eq. (10.3) is not so easy to justify. Indeed if we were to reverse the subscripts m and n in Eq. (10.3) to write

$$I_m = \left(\frac{1}{q}\right) \sum_n G_{n,m} (\mu_m - \mu_n) \longrightarrow WRONG!$$

it would not even be correct. Its predictions would be different from those of Eq. (10.3) for multi-terminal non-reciprocal circuits (Fig. 10.11).

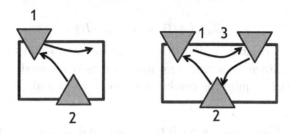

Fig. 10.11 A magnetic field makes an electron coming in from contact 2 veer towards contact 1, but makes an electron coming from contact 1 veer away from contact 2. Is $G_{1,2} \neq G_{2,1}$? Yes, if there are more than two terminals, but not in a two-terminal circuit.

10.3.3 *Non-reciprocal circuits*

This may be a good place to raise an interesting property of conductors with non-reciprocal transmission of the type expected from edge states. Consider the structure shown in Fig. 10.11 with a B-field that makes an electron coming in from contact 2 veer towards contact 1, but makes an electron coming from contact 1 veer away from contact 2. Is $G_{1,2} \neq G_{2,1}$?

The answer is "no" in the linear response regime as evident from the sum rule (Eq. (10.13)) which for a structure with two terminals requires that

$$G_{1,1} = G_{1,2} = G_{2,1}.$$

However, there is no such requirement for a structure with more than two terminals. For example with three terminals, Eq. (10.13) tells us that

$$G_{1,1} = G_{1,2} + G_{1,3} = G_{2,1} + G_{3,1}$$

which does not require $G_{1,2}$ to equal $G_{2,1}$.

The effects of such non-reciprocal transmission have been observed clearly with "edge states" in the quantum Hall regime (discussed at the end of Chapter 11). This idea of "edge states" providing unidirectional ballistic channels over macroscopic distances is a very remarkable effect, but it has so far been restricted to low temperatures and high B-fields making it not too relevant from an applied point of view. That may change with the advent of new materials like "topological insulators" which show edge states even without B-fields.

But can we have non-reciprocal transmission without magnetic fields? In general the conductance matrix (which is proportional to the transmission matrix) obeys the *Onsager reciprocity relation* (see Section 10.3.4

below)

$$G_{n,m}(+B) \; = \; G_{m,n}(-B) \tag{10.26}$$

requiring the current at n due to a voltage at m to equal the current at m due to a voltage at n with any magnetic field reversed. Doesn't this Onsager relation require the conductance to be reciprocal

$$G_{n,m} \; = \; G_{m,n}$$

when $B = 0$? The answer is yes if the structure does not include magnetic materials. Otherwise we need to reverse not just the external magnetic field but the internal magnetization too.

$$G_{n,m}(+B,+M) \; = \; G_{m,n}(-B,-M). \tag{10.27}$$

For example if one contact is magnetic, Onsager relations would require the $G_{1,2}$ in structure (a) to equal $G_{2,1}$ in structure (b) with the contact magnetization reversed as sketched above. *But that does not mean $G_{1,2}$ equals $G_{2,1}$ in the same structure,* (a) or (b).

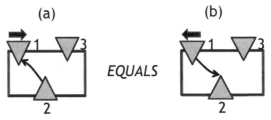

And so based on our current understanding a "topological insulator" which is a non-magnetic material could not show non-reciprocal conductances at zero magnetic field with ordinary contacts, but might do so if magnetic contacts were used. But this is an evolving story whose ending is not yet clear.

10.3.4 *Onsager relations*

Before moving on, let me quickly outline how we obtain the Onsager relations (Eq. (10.26)) requiring the current at 'n' due to a voltage at 'm' to be equal to the current at 'm' due to a voltage at 'n' with any magnetic field reversed. This is usually proved starting from the multi-terminal version of the Kubo formula (Chapter 5)

$$G_{m,n} = \frac{1}{2kT} \int_{-\infty}^{+\infty} d\tau \langle I_m(t_0 + \tau) I_n(t_0) \rangle_{eq} \tag{10.28}$$

involving the correlation between the currents at two different terminals.

Consider a three terminal structure with a magnetic field $(B > 0)$ that makes electrons entering contact 1 bend towards 2, those entering 2 bend towards 3 and those entering 3 bend towards 1.

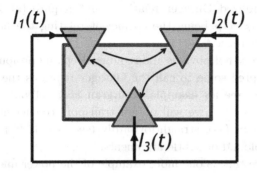

We would expect the correlation

$$\langle I_2(t_0 + \tau)I_1(t_0)\rangle_{eq}$$

to look something like this sketch with the correlation extending further for positive τ. This is because electrons go from 1 to 2, and so the current I_1 at time t_0 is strongly correlated to the current I_2 at a later time $(\tau > 0)$, but not to the current at an earlier time.

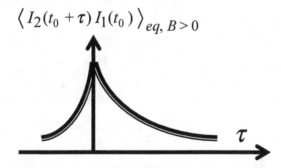

If we reverse the magnetic field $(B < 0)$, it is argued that the trajectories of electrons are reversed, so that

$$\langle I_1(t_0 + \tau)I_2(t_0)\rangle_{eq, \ B<0} = \langle I_2(t_0 + \tau)I_1(t_0)\rangle_{eq, \ B>0}. \tag{10.29}$$

This is the key argument. If we accept this, the Onsager relation (Eq. (10.26)) follows readily from the Kubo formula (Eq. (10.28)).

What we have discussed here is really the simplest of the Onsager relations for the generalized transport coefficients relating generalized forces to fluxes. For example, in Chapter 13 we will discuss additional coefficients like G_S relating a temperature difference to the electrical current. There are generalized Onsager relations that require (at zero magnetic field) $G_P = TG_S$, G_P being the coefficient relating the heat current to the potential difference.

This is of course not obvious and requires deep and profound arguments that have prompted some to call the Onsager relations the fourth law of thermodynamics (see for example, Yourgrau *et al.* 1966). Interestingly, however, in Chapter 13 we will obtain transport coefficients that satisfy this relation $G_P = TG_S$ straightforwardly (see Eqs. (13.6) and (13.17a)) without any profound or subtle arguments.

We could cite this as one more example of the power and simplicity of the elastic resistor that comes from disentangling mechanics from thermodynamics.

Chapter 11

Hall Effect

11.1 Introduction

Let me briefly explain what the Hall effect is about. Consider a two-dimensional conductor (see Fig. 11.1) carrying current, subject to a perpendicular magnetic field along the y-direction which exerts a force on the electrons perpendicular to its velocity.

$$\mathbf{F} = \frac{d\mathbf{p}}{dt} = -q\mathbf{v} \times \mathbf{B}. \tag{11.1}$$

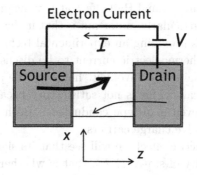

Fig. 11.1 A magnetic field B in the y-direction makes electrons from the source veer "up"wards.

This would cause an electron from the source to veer "up"wards and an electron from the drain to veer "down"wards as shown. Since there are more electrons headed from source to drain, we expect electrons to pile up on the top side causing a voltage V_H to develop in the x-direction transverse to current flow (see Fig. 11.2).

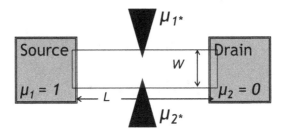

Fig. 11.2 Basic structure with two voltage probes whose potential difference measure the Hall voltage, $qV_H = \mu_1^* - \mu_2^*$.

The Hall effect has always been important since its discovery around 1880, and has acquired a renewed importance since the discovery of the quantum Hall effect in 1980 at high magnetic fields. In this book we have seen the conductance quantum q^2/h appear repeatedly and it is very common in the context of nanoelectronics and mesoscopic physics. But the quantum Hall effect was probably the first experimental observation where it played a clear identifiable role and the degree of precision is so fantastic that the National Bureau of Standards uses it as the resistance standard. We will talk briefly about it later at the end of this chapter. For the moment let us focus on the conventional Hall effect at low magnetic fields.

One reason it is particularly important is that it changes sign for n- and p-type materials, thus providing an experimental technique for telling the difference. Like the thermoelectric current to be discussed in Chapter 13, this is commonly explained by invoking "holes" as the positive charge carriers in p-type materials. This is not satisfactory because it is really the electrons that move even in p-type conductors. Both n-type and p-type conductors have negative charge carriers.

For the thermoelectric effect we will see that its sign is determined by the slope of the density of states $D(E)$, that is whether it is an increasing or a decreasing function of the energy E. By contrast, the sign of the Hall effect is determined by the sign of the effective mass defined as the ratio of the momentum p to the velocity dE/dp (see Chapter 6). As a result although the magnetic force (see Eq. (11.1)) is the same for both n- and p-type conductors, giving the same $d\mathbf{p}/dt$, the resulting $d\mathbf{v}/dt$ has opposite signs. This makes electrons in p-conductors veer in the opposite direction giving rise to a Hall voltage of the opposite sign. Clearly this requires the existence of an $E(p)$ relation underlying the density of states function.

Perhaps it is for this reason that amorphous semiconductors which lack a well-defined $E(p)$ often show strange results for the Hall effect and yet show reasonable thermoelectric effect.

The simple theory of the Hall effect given in freshman physics texts goes like this. First the current is written as

$$I = q(N/L)v_d \qquad (11.2)$$

with the drift velocity given by the product of the mobility and the electric field in the z-direction:

$$v_d = \bar{\mu} \, (V/L). \qquad (11.3)$$

These two relations are normally combined to yield the Drude formula (see Eq. (6.1))

$$\frac{I}{V} = \underbrace{q \frac{N}{WL} \bar{\mu}}_{\sigma} \frac{W}{L}. \qquad (11.4)$$

For the Hall effect, it is argued that an electric field V_H/W must appear in the x-direction to offset the magnetic force

$$\frac{V_H}{W} = v_d B. \qquad (11.5)$$

Combining Eq. (11.5) with Eq. (11.2) one obtains the standard expression for the Hall resistance

$$R_H = \frac{V_H}{I} = \frac{B}{q(N/LW)}. \qquad (11.6)$$

One reason the Hall effect is widely used is that Eq. (11.6) allows us to determine the electron density N/LW from the slope of the Hall resistance versus B-field curve. This looks like a straightforward transparent theory for a well-established effect. What more could we add to it?

The main concern we have about this derivation is the same concern that we voiced regarding the Drude formula, namely that if electric field were indeed what drives currents then all electrons should feel its effect. Indeed Eq. (11.6) for the Hall resistance conveys the impression that the Hall effect depends on the total electron density N/LW over all energies. But we believe this is not correct.

Like the other transport coefficients we have discussed, the Hall resistance too is a "Fermi surface property" that depends only on the electrons in an energy window $\sim$ a few kT around $E = \mu_0$ and not on the total number of electrons obtained by integrating over energy. We will show that the Hall resistance for a single energy channel of an elastic resistor is given by

$$R_H(E) = \frac{2BLW}{qD(E)v(E)p(E)} \qquad (11.7)$$

which can be averaged over an energy window $\sim$ of a few kT around $E = \mu_0$ using our standard broadening function:

$$\frac{1}{R_H} = \int_{-\infty}^{+\infty} dE \left(-\frac{\partial f_0}{\partial E} \right) \frac{1}{R_H(E)}. \qquad (11.8)$$

Note that in general we should integrate the conductance $1/R_H$ rather than the resistance R_H since different energy channels all have the same voltage so that they conduct "in parallel" as circuit experts would put it. Equations (11.8) and (11.7) can be reduced to the standard result (Eq. (11.6)) by making use of the single band relation obtained in Chapter 6

$$D(E)v(E)p(E) = N(E) \cdot d \quad \text{(same as Eq. (6.9))}$$

with $d = 2$ for a two-dimensional conductor and relating the average of $N(E)$ to the total number of electrons as we did in Section 6.5.

But if the single band relation (Eq. (6.9)) is not applicable one should use the expression in Eq. (11.8) rather than Eq. (11.6). In any case, Eq. (11.8) shows that the effect really does not involve electrons at all energies. One reason this point causes some confusion is the existence of equilibrium currents inside the sample in the presence of a magnetic field which involve all electrons at all energies.

However, these are dissipationless currents of the type that exist even if we put a hydrogen atom in a magnetic field and have nothing to do with the transport coefficients we are talking about. In any transport model it

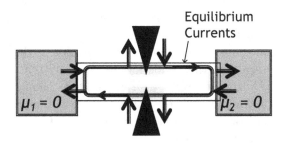

Fig. 11.3   Equilibrium currents can exist in any conductor in the presence of a magnetic field.

is important to eliminate these non Fermi surface currents. A similar issue arises with respect to spin currents even without any magnetic fields (which we will discuss in Part B).

Getting back to the problem of determining the Hall voltage, as we saw in Chapter 10 there are two approaches: (a) calculate the non-equilibrium electrochemical potential inside the conductor or (b) treat it as a four terminal structure using the Büttiker equation. We will not discuss the second one here since the basic idea was already discussed at the end of Chapter 10. Instead we will focus on the first approach which gives insights into the variation of the electrochemical potential inside the channel. But first let us briefly discuss the dynamics of electrons in a B-field.

11.2 Why n- and p-type Conductors are Different

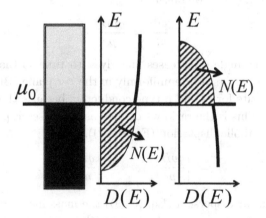

Fig. 11.4 The Hall resistance changes sign for n- and p-type conductors and is inversely proportional to $N(E)$ at $E = \mu_0$.

Why do n- and p-type conductors show opposite signs of the Hall effect? The basic difference is that in n-type conductors, the velocity is parallel to the momentum, while in p-type conductors, it is anti-parallel because $v = dE/dp$, and in p-type conductors the energy decreases with increasing p (Fig. 11.4). To see why the relative sign of p and v matters, let us consider the magnetic force described by Eq. (11.1) in a little more detail.

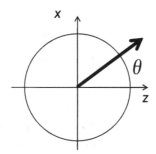

For any isotropic $E(p)$ relation, the velocity and momentum are collinear (parallel or anti-parallel) pointing, say at an angle θ to the z-axis, so that

$$\mathbf{p} = p\cos\theta\,\hat{\mathbf{z}} + p\sin\theta\,\hat{\mathbf{x}}$$

$$\mathbf{v} = v\cos\theta\,\hat{\mathbf{z}} + v\sin\theta\,\hat{\mathbf{x}}.$$

Inserting into Eq. (11.1) we obtain

$$\frac{d\theta}{dt} = \frac{qvB}{p} \tag{11.9}$$

showing that the angle θ increases linearly with time so that the velocity and momentum vectors rotate uniformly in the z-x plane. But the sense of rotation is opposite for n- and p-type conductors because the ratio p/v has opposite signs. This is the ratio we defined as mass (see Eq. (6.11)) and is constant for parabolic dispersion (Eq. (6.12)).

$$\omega_c = \left|\frac{qvB}{p}\right|_{E=\mu_0} = \left|\frac{qB}{m}\right|_{E=\mu_0}. \tag{11.10}$$

But for a linear dispersion (Eq. (6.13)) the mass increases with energy, so that the cyclotron frequency decreases with increased carrier density, as is observed in graphene. The magnetic field tries to make the electron go round and round in a circle with an angular frequency ω_c. However, it gets scattered after a mean free time τ, so that if $\omega_c\tau \ll 1$ the electron never really gets to complete a full rotation. This is the low field regime where the Hall resistance in given by Eq. (11.6), while the high field regime characterized by $\omega_c\tau \gg 1$ leads to the quantum Hall effect mentioned earlier. Let us now discuss our first approach to the problem of determining the Hall resistance (Eq. (11.7)) based on looking at the non-equilibrium electrochemical potentials inside the conductor.

11.3   Spatial Profile of Electrochemical Potential

As I mentioned earlier, the textbook derivation of the Hall resistance (Eq. (11.6)) looks fairly straightforward, but we are attempting to provide a different expression (Eq. (11.7)) motivated by the same reasons that prompted us to describe an alternative expression for the conductivity back in Chapter 6.

In our elastic resistor model, the role of the drift velocity in the textbook derivation is played by the potential separation

$$\delta\mu \ = \ \mu^+ \ - \ \mu^-$$

between drainbound and sourcebound states, so that instead of Eq. (11.2) we have (see Eq. (8.23))

$$I(E) \ = \ \frac{q}{h} M(E) \left(-\frac{\partial f_0}{\partial E}\right) \delta\mu \tag{11.11}$$

$$\text{with} \quad \frac{M(E)}{h} \ = \ \frac{D(E)v(E)}{\pi L} \tag{11.12}$$

where we have used the result for 2D conductors from Eq. (4.13).

Instead of Eq. (11.3), we have the potential separation related to the applied voltage by

$$\delta\mu \ = \ \frac{qV\lambda}{L+\lambda} \ \cong \ qV\frac{\lambda}{L}. \tag{11.13}$$

This relation can be obtained by noting that $\delta\mu$ is equal to the fraction of the applied qV that is dropped across the interface resistance R_B (see Fig. 8.2):

$$\delta\mu \ = \ qV\frac{R_B}{R} \ = \ qV\frac{G}{G_B}.$$

Making use of Eq. (4.5a) we obtain Eq. (11.13).

Just as Eqs. (11.2) and (11.3) yield the Drude expression for the conductivity, similarly Eqs. (11.11) and (11.13) can be combined to yield the more general conductivity expression discussed earlier (see Eq. (6.5)).

For the Hall effect we also need a replacement for Eq. (11.5)

$$\frac{V_H}{W} \ = \ v_d\, B$$

which we will show is given by

$$\frac{V_H}{W} \ = \ \frac{2}{\pi}\frac{\delta\mu}{p}B \tag{11.14}$$

Eq. (11.14) together with Eqs. (11.11) and (11.13) gives us the result for Hall resistance stated earlier in Eq. (11.7). Unfortunately we do not have a one-line argument for Eq. (11.14) like the one used for Eq. (11.5). Instead I need to describe a two-page argument using the BTE discussed in Chapter 9.

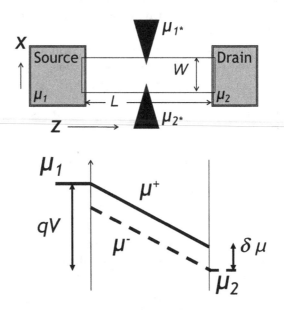

Fig. 11.5   Spatial variation of $\mu^{\pm}$ along z.

11.3.1   *Obtaining Eq. (11.14) from BTE*

Back in Chapter 9 we obtained a solution for a subset of this problem based on a solution of

$$v_z \frac{\partial \mu}{\partial z} = -\frac{\mu - \overline{\mu}}{\tau} \tag{11.15}$$

and obtained the solutions for the electrochemical potentials μ^{+} and μ^{-} sketched above in Fig. 11.5. The solutions could be written in the form

$$\mu(z, \theta) = \overline{\mu}(z) + \frac{2}{\pi} \delta\mu \, \cos\theta \tag{11.16}$$

where we have separated out a z-dependent part $\overline{\mu}$ from the momentum-dependent part at a specific location, z. The latter needs a little discussion.

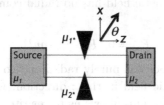

Since we are discussing an elastic resistor for which electrons have a fixed energy E and hence a fixed momentum p, it is convenient to use cylindrical coordinates for the momentum (p, θ) instead of (p_x, p_z). Suppose we only had electrons traveling along θ and its exact reverse direction. Making use of Eq. (11.13) we could write

$$\mu(z) = \overline{\mu}(z) + \frac{\delta\mu}{2} = \overline{\mu}(z) + \frac{qV}{2L}\overline{\lambda}(\theta). \tag{11.17}$$

Since we only have electrons along a single direction θ, we can write

$$\overline{\lambda}(\theta) = 2\nu_z\tau = 2\nu\tau \cos\theta$$

so that from Eq. (11.17)

$$\mu(z) = \overline{\mu}(z) + \frac{qV}{L}\nu\tau \cos\theta \tag{11.18}$$

Eq. (11.16) follows on using Eq. (11.13) again, but this time using the usual θ-independent mean free path is given in 2D by Eq. (4.10)

$$\lambda = \frac{\pi}{2}\nu\tau.$$

The question is how we expect the solution in Eq. (11.16) to change when we turn on the magnetic field so that it exerts a force on the electrons. For this we could use a linearized version of the BTE like Eq. (9.17), but retaining both z- and x-components since we have a two-dimensional problem

$$v_x\frac{\partial\mu}{\partial x} + v_z\frac{\partial\mu}{\partial z} + F_x\frac{\partial\mu}{\partial p_x} + F_z\frac{\partial\mu}{\partial p_z} = -\frac{\mu - \overline{\mu}}{\tau}. \tag{11.19}$$

Note that Eq. (11.15) is a "subset" of this equation which includes three extra terms. The last two coming from the magnetic force (Eq. (11.1)) can be written as

$$F_x\frac{\partial\mu}{\partial p_x} + F_z\frac{\partial\mu}{\partial p_z} = \mathbf{F} \cdot \nabla_{\mathbf{p}}\mu = \frac{F_\theta}{p}\frac{\partial\mu}{\partial\theta} + F_r\frac{\partial\mu}{\partial p}.$$

The force due to a magnetic field has no radial component, only a θ component:

$$F_r = 0, \ F_\theta = -qvB.$$

This is because the velocity is purely radial and so when we take a cross-product with a magnetic field in the z-direction, we get a vector that is purely in the θ-direction. This allows us to rewrite Eq. (11.19) in the form

$$v_x \frac{\partial \mu}{\partial x} + v_z \frac{\partial \mu}{\partial z} - \frac{qvB}{p} \frac{\partial \mu}{\partial \theta} = -\frac{\mu - \bar{\mu}}{\tau}. \tag{11.20}$$

Noting that our solution in Eq. (11.16) satisfies Eq. (11.15), it is easy to check that if we add an extra term varying only with x to it, the resulting expression

$$\mu(z, \theta, x) = \bar{\mu}(z) + \frac{2}{\pi} \delta\mu \, \cos\theta - \frac{2}{\pi} \frac{\delta\mu}{p} \, qBx \tag{11.21}$$

will satisfy Eq. (11.20). From this solution we obtain the desired result in Eq. (11.14) by writing

$$-qV_H = \mu(x = W) - \mu(x = 0) = -\frac{2}{\pi} \frac{\delta\mu}{p} qBW.$$

11.4 Edge States

As we mentioned in the introduction, a very important discovery is the quantum Hall effect observed when the B-fields are so high that electrons from the source "hug" one edge, while electrons from the drain hug the other edge of the sample due to the formation of the so-called "skipping orbits" (Fig. 11.6)

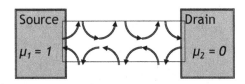

Fig. 11.6 Skipping orbits in high B-fields leads to a "divided highway" with drainbound electrons on one side and sourcebound electrons on the other.

Under these conditions one edge of the channel is at an electrochemical potential equal to that at the source, while the other edge is at a potential equal to the drain, so that the potential drop across the width (or the

Hall voltage) is equal to that between the source and the drain. This is very different from the ordinary Hall effect when the Hall voltage given by Eq. (11.14) is a small fraction of the applied voltage. What makes the quantum Hall effect so extraordinary is that the Hall resistance (Hall voltage divided by the current) is given by

$$R = \frac{h}{q^2 i} \tag{11.22}$$

where i is an integer to a fantastic degree of precision, making this a resistance standard used by the National Bureau of Standards. It is as if we have an unbelievably perfect ballistic conductor whose only resistance is the interface resistance.

Since these conductors are often hundreds of micrometers long, this perfect ballisticity is amazing and was recognized with a Nobel prize in 1985 (von Klitzing K. *et al.* 1980). One can ascribe this incredible ballisticity to the formation of a "divided" electronic highway (Fig. 11.6) with drainbound electrons so well-separated from the sourcebound electrons that backscattering is extremely unlikely (Fig. 11.6). This simple picture, however, is a little too simple. It does not for example tell us the significance of the integer i in Eq. (11.22) which requires some input from quantum mechanics, as we will see in Part B.

Chapter 12

Smart Contacts

A key insight of modern nanoelectronics is the concept of an interface resistance R_B that is in series with the standard length-dependent resistance Eq. (1.8):

$$R = R_B\left(1 + \frac{L}{\lambda}\right).$$

The interface resistance depends solely on the properties of the channel and cannot be eliminated even with an ideal contact.

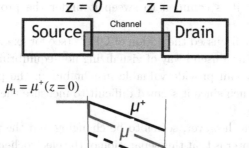

Fig. 12.1 Spatial profile of electrochemical potentials μ^+ and μ^- across a diffusive channel. (Same as Fig. 8.3)

As we saw in Chapter 8, the key concept in identifying this interface resistance was the recognition that when a current flows, the electrochemical potentials μ^+ and μ^- for the drainbound and sourcebound states are different (Fig. 12.1). From Eq. (11.13) we could write (Note: $\mu_1 - \mu_2 = qV$)

$$\delta\mu \equiv \mu^+ - \mu^- = \frac{\mu_1 - \mu_2}{1 + L/\lambda}. \tag{12.1}$$

The contacts held at different potentials μ_1 and μ_2 drive the two groups of states (drainbound and sourcebound) out of equilibrium, while backscattering processes described by the mean free path λ try to restore equilibrium. Equation (12.1) describes the result of these competing forces.

Normally we do not like to deal with multiple electrochemical potentials. The diffusion equation for example (see Eq. (7.17)),

$$\frac{I}{A} = -\frac{\sigma_0(z)}{q}\frac{d\mu}{dz} \tag{12.2}$$

works in terms of a single potential $\mu(z)$ and what we saw in Chapter 8 was how we could avoid talking about the two potentials $\mu^+(z)$ and $\mu^-(z)$ by defining $\mu(z)$ as the average of the two and including interface resistances into the boundary conditions by replacing Eq. (8.3) with Eq. (8.6).

Non-equilibrium flow of electrons requires two contacts with separate electrochemical potentials μ_1 and μ_2 in the two contacts, and a spatially varying $\mu(z)$ in between. But many feel uncomfortable with the notion of multiple electrochemical potentials or Quasi-Fermi Levels (QFL's) inside the channel and it is common to sweep it under the proverbial rug, as described above.

I have always discussed the notion of QFL's (see for example, Chapter 2 of Datta (1995)) as a useful way of visualizing non-equilibrium states inside the channel that can provide valuable insight but in the past I have not stressed it too much since it seemed difficult to measure the QFL's μ^+ and μ^- shown in Fig. 12.1.

This situation, however, seems to be changing and the point I wish to make in this chapter is that this separation of the electrochemical potentials for different groups of states is really far more ubiquitous and cannot always be swept under the rug. Indeed I would like to go further and argue that the most interesting devices of the future could well be the ones where multiple electrochemical potentials will represent *the essential physics*.

Let me present two examples. The first is an old example from an old device, the p-n junction for which the need for two separate electrochemical potentials for the conduction and valence bands is well-recognized. The second is a more recent example from the fast-developing field of spintronics where distinct QFL's for up and down spins are becoming quite common aided by the development of magnetic contacts that can be used to measure the spin QFL's.

12.1   p-n Junctions

Figure 12.2 shows a grayscale plot of the density of states $D(z, E)$. The white band indicates the bandgap with a non-zero DOS both above and below it on each side which are shifted in energy with respect to each other. A positive voltage is applied to the right with respect to the left, so that μ_2 is lower than μ_1 as shown.

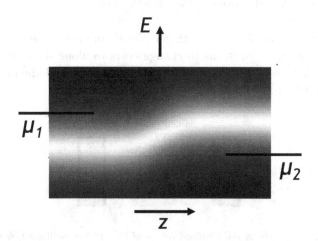

Fig. 12.2   Simplified grayscale plot of the spatially varying density of states $D(z, E)$ across a p-n junction.

If we look at a narrow range of energies around μ_1 it communicates primarily with contact 1. If we look at a narrow range of energies around μ_2 it communicates primarily with contact 2. We could draw an *idealized* diagram with each of these two groups communicating just with one contact and cut off from the other as shown in Fig. 12.3. In reality of course neither group is completely cutoff from either contact, and people who design real devices often go to great lengths to achieve better isolation, but let us not worry about such details.

Would the idealized device in Fig. 12.3 allow any current to flow? None at all, if it were an elastic resistor. There is no energy channel that will let an electron get all the way from left to right. The ones connected to the left are disconnected from the right and those connected to the right are disconnected from the left.

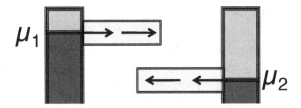

Fig. 12.3 An idealized version of the p-n junction in Fig. 12.2, assuming complete isolation of each channel from one of the contacts.

But current can flow even with such ideal contacts because of inelastic processes that allow electrons to change energies along the channel. Electrons can then come in from the left, change energy and then exit to the right as sketched in Fig. 12.4.

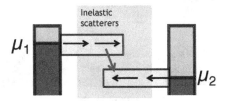

Fig. 12.4 Current flow in the idealized device of Fig. 12.3 is facilitated by distributed inelastic processes.

Indeed this is exactly how currents flow in long p-n junctions, by transferring from the upper group of states down to the lower group by inelastic processes, which are generally referred to as recombination-generation (R-G) processes, since people like to think in terms of electrons in the upper group recombining with a "hole" in the lower group. But as we mentioned in Chapter 6 this is really an unnecessary complication and one could simply think purely in terms of electrons transferring inelastically from one group of states to another.

The point to note is that this class of devices cannot be described with one electrochemical potential and to capture the correct physics, it is essential to treat the two groups of states separately, *introducing two different electrochemical potentials*, labeled with the index n

$$\frac{I_n}{A} = -\frac{\sigma_{0,n}(z)}{q}\frac{d\mu_n}{dz}. \tag{12.3}$$

These currents are all coupled together by inelastic processes generally called recombination-generation or "RG" processes in the context of p-n junctions

$$\frac{dI_n}{dz} = \sum_m \left(RG_{m \to n} - RG_{n \to m} \right) \tag{12.4}$$

that take electrons from one group of states m to the other n. This is indeed the way p-n junctions are modeled.

It is well-known that the current in a p-n junction is given by an expression of the form

$$I = I_0 \left(e^{qV/\nu kT} - 1 \right) \tag{12.5}$$

where the number ν as well as the coefficient I_0 are determined by the nature of the inelastic or RG processes. The conductivities of either of the two groups of states play hardly any role in determining this current.

The physical reason for this is clear. The rate-determining step in current flow is the inelastic process transferring electrons from one group of states to the other. Transport within any of these groups only adds an unimportant resistance in series with the basic device.

Everything we have talked about in this book has been about the conductivities σ_n of the homogeneous p-type or n-type materials. And this is exactly the physics that is relevant to the operation of the most popular electronic device today, namely the Field Effect Transistor (FET) whose conductivity is controlled by a gate electrode through the electrostatic potential U.

But the p-n junction is a totally different device from the FET both in terms of its current-voltage characteristics and the physics that underlies it. It is the basic device structure used to construct solar cells and the principle it embodies is key to a broad class of energy conversion devices. So let me take a short detour to elaborate on this principle.

12.1.1 *Current-voltage characteristics*

Consider for example the device in Fig. 12.5 assuming that the upper group of states (labeled A) is clustered around an energy ε_A while the lower group (labeled B) is clustered around ε_B.

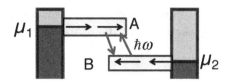

Fig. 12.5   Same as Fig. 12.4 with the two groups of states labeled A and B. Electronic transitions between A and B are facilitated by inelastic interactions.

The essential physics of such p-n junction like devices is contained not in Eq. (12.3), but in Eq. (12.4) which for two levels A and B can be written as

$$I \sim D_{BA} f_A(\varepsilon_A)\left(1 - f_B(\varepsilon_B)\right) - D_{AB} f_B(\varepsilon_B)\left(1 - f_A(\varepsilon_A)\right) \qquad (12.6)$$

where the coefficients D_{BA} and D_{AB} denote the strength of the inelastic processes inducing the transitions from A to B and from B to A respectively (note that the transition occurs from the second subscript to the first).

Interestingly these two rates D_{AB} and D_{BA} are generally NOT equal. D_{AB} involves absorbing an amount of energy

$$\hbar\omega = \varepsilon_A - \varepsilon_B$$

from the surroundings, while D_{BA} involves giving up the same amount of energy.

A fundamental principle of equilibrium statistical mechanics (see Chapter 15) is that if the entity causing the inelastic scattering is at equilibrium with a temperature T_0, then it is always harder to absorb energy from it than it is to give up energy to it and the ratio of the two processes is given by

$$\frac{D_{AB}}{D_{BA}} = \exp\left(-\frac{\hbar\omega}{kT_0}\right). \qquad (12.7)$$

We can write the current from Eq. (12.6) in the form

$$I \sim D_{AB} f_B(\varepsilon_B)\left(1 - f_A(\varepsilon_A)\right)\left(X - 1\right) \qquad (12.8)$$

$$\text{where} \quad X \equiv \frac{D_{BA}}{D_{AB}} \frac{f_A(\varepsilon_A)}{1 - f_A(\varepsilon_A)} \frac{1 - f_B(\varepsilon_B)}{f_B(\varepsilon_B)}. \qquad (12.9)$$

Making use of Eqs. (12.8) and (12.9) and the following property of Fermi functions (Eq. (2.2))

$$\frac{1 - f_0(\varepsilon)}{f_0(\varepsilon)} = \exp\left(\frac{\varepsilon - \mu_0}{kT}\right) \tag{12.10}$$

we can rewrite Eq. (12.9) as

$$X = \exp\left(\frac{\hbar\omega}{kT_0} - \frac{\hbar\omega}{kT}\right) \exp\left(\frac{\mu_A - \mu_B}{kT}\right). \tag{12.11}$$

Since Level A is connected to contact 1 and Level B to contact 2, if the inelastic processes taking electrons from A to B are not too strong, level A is almost in equilibrium with contact 1 and level B with contact 2, so that

$$\mu_A - \mu_B \approx \mu_1 - \mu_2 = qV.$$

If $T_0 = T$, we can write the current from Eq. (12.8) as

$$I \sim (X - 1) \sim e^{qV/kT} - 1$$

which is the ideal *I-V* relation for p-n junctions stated earlier (see Eq. (12.5)) with $\nu = 1$. Other values of ν are also obtained in practice but that requires a more detailed discussion beyond the scope of this book.

The more important point I want to stress is that this device can be used for *energy conversion*. If the scatterers are at a temperature different from that of the device ($T_0 \neq T$) then one can have a current flowing even without any applied voltage. This short circuit current is given by

$$I_{sc} \equiv I(V = 0) \sim \exp\left\{\frac{\hbar\omega}{k}\left(\frac{1}{T_0} - \frac{1}{T}\right)\right\} - 1. \tag{12.12}$$

One could in principle use a device like this to convert a temperature difference ($T_0 \neq T$) into an electrical current. The short circuit current has the opposite sign for $T_0 > T$ and for $T > T_0$. Readers familiar with Feynman's ratchet and pawl lecture (Feynman 1963, cited in Chapter 15) may notice the similarity. The ratchet reverses direction depending on whether its temperature is lower or higher than the ambient.

One could view more practical devices like solar cells as embodiments of the same principle, the light from the sun having a temperature $T_0 \sim 6000°C$ characteristic of the surface of the sun, much larger than the ambient temperature.

From Eq. (12.8) it is easy to see that under open circuit conditions ($I = 0$), we must have $X = 1$, so that from Eq. (12.11) we have

$$\frac{qV_{OC}}{\hbar\omega} = 1 - \frac{T}{T_0}.$$

The left hand side represents the energy extracted per photon under very low current (near open circuit) conditions, so that this could be called the Carnot efficiency of a solar cell viewed as a "heat engine". However, since $T_0 \gg T$, this Carnot efficiency is very close to 100% and my colleague Ashraf often points out that other factors related to the small angular spectrum of solar energy are important in lowering the ideal efficiency to much lower values.

12.1.2 *Contacts are fundamental*

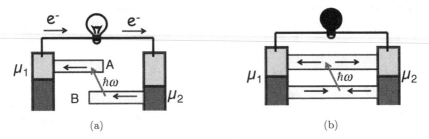

(a)   (b)

Fig. 12.6   (a) Asymmetric contacts are central to the operation of the "solar cell". (b) If contacted symmetrically no electrical output is obtained.

The point I want to make is how important the discriminating contacts are in the design of this class of devices which we could generally refer to as "solar cells" (Fig. 12.6a). The external source raises electrons from the B states to the A states from where they exit through the left contact, while the empty state left behind in B is filled up by an electron that comes in through the right contact. Every electron raised from B to A thus causes an electron to flow in the external circuit.

But if the contacts are connected "normally" injecting and extracting equally from either group (Fig. 12.6b) then we cannot expect any current to flow in the external circuit, from the sheer symmetry of the arrangement. After all, why should electrons flow from left to right any more that they would flow from right to left?

It is this asymmetric contacting that makes p-n junctions fundamentally different from the Field Effect Transistor (FET) that we started our chapters with, both in terms of the current-voltage characteristics and the physics underlying it. It is of course well recognized that the physics of p-n junctions demands two different electrochemical potentials. What is not as well recognized is the generic nature of this phenomenon. Let us now look at a more recent example.

12.2 Spin Potentials

Related video lecture available at course website, Unit 3: L3.9.

12.2.1 *Spin valve*

One of the major developments in the last two decades is the spin valve, a device with two magnetic contacts (Fig. 12.7). If they are magnetized in the same direction (parallel configuration, P) the resulting resistance is lower than if they are magnetized in opposite directions (anti-parallel configuration, AP). Since its first demonstration in 1988, it rapidly found application as a "reading" device to sense the information stored in a magnetic memory and the discovery was recognized with a Nobel prize in 2007.

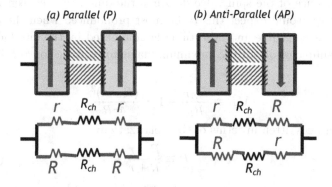

Fig. 12.7 Spin valve: (a) Parallel (P) configuration. (b) Anti-parallel (AP) configuration.

So far we have only mentioned spin as part of a "degeneracy factor, g" (Section 6.4.1), the idea being that electronic states always come in pairs, one corresponding to each spin. We could call these "up" and "down" or "left" and "right" or even "red" and "blue" as we have done in Fig. (12.7). Note that the two spins are not spatially separated even though we have separated the red and the blue channel for clarity. Ordinarily, the two channels are identical and we can calculate the conductance due to one and remember to multiply by two.

But in spin valve devices the contacts are magnets that treat the two spin channels differently and the operation of a spin valve can be understood in fairly simple terms if we postulate that each spin channel has a different interface resistance with the magnet depending on whether it is parallel (majority spin) or anti-parallel (minority spin) to the magnetization.

If we assume the interface resistance for majority spins to be r and for minority spins to be R ($r < R$) we can draw simple circuit representations for the P and AP configurations as shown, with R_{ch} representing the channel resistance. Elementary circuit theory then gives us the resistance for the parallel configuration as

$$R_P = \left(\frac{1}{2r + R_{ch}} + \frac{1}{2R + R_{ch}} \right)^{-1} \tag{12.13}$$

and that for the anti-parallel configuration as

$$R_{AP} = \frac{r + R + R_{ch}}{2}. \tag{12.14}$$

The essence of the spin valve device is the difference between R_P and R_{AP} and we would expect this to be most pronounced when the channel resistance is negligible and everything is dominated by the interfaces. We obtain a simple result for the maximum magnetoresistance or MR if we set $R_{ch} = 0$

$$MR \equiv \frac{R_{AP}}{R_P} - 1 = \frac{(R - r)^2}{4rR} \tag{12.15}$$

which can be written in terms of the polarization:

$$P \equiv \frac{R - r}{R + r} \tag{12.16a}$$

$$MR = \frac{P^2}{1 - P^2}. \tag{12.16b}$$

I should mention here that the expression commonly seen in the literature has an extra factor of 2 compared to Eq. (12.16b)

$$MR = \frac{2P^2}{1 - P^2}$$

which is applicable to magnetic tunnel junctions (MTJ's) that use short tunnel junctions as channels instead of the metallic channels we have been discussing. We get this extra factor of 2, if we assume that two resistors R_1 and R_2 in series give a total resistance of $K R_1 R_2$, K being a constant, instead of the standard result $R_1 + R_2$ expected of ordinary Ohmic resistors. The product dependence captures the physics of tunnel resistors.

While spin valves showed us how to use magnets to inject spins and control spin potentials, later researchers have shown how to use non-equilibrium spins to turn nanoscale magnets thus integrating spintronics and magnetics into a single and very active area of research with exciting possibilities for which the reader may want to look at some of the current literature.

Our objective is simply to point out the existence of different internal potentials for different spins. The key to spin valve operation is the different interface resistances, r and R, associated with each spin for magnetic contacts as shown in the simple circuit in Fig. 12.7. The same circuit also shows that the potential profile will be different for the two spin channels, since each channel has a different set of resistances.

It is now well established that magnetic contacts can generate different spin potentials, but we will not discuss this further. Instead let me end by talking about a recent discovery showing that spin potentials can be generated even without magnetic contacts in channels with high spin-orbit coupling.

There is great current interest in a new class of materials called topological insulators where the electronic eigenstates at a surface exhibit "spin-momentum locking" such that their spin is perpendicular to their momentum, and their cross-product is in the direction of the surface normal. What this means is that the QFL's μ^+ and μ^- shown in Fig. 12.8 translate into spin QFL's μ_{up} and μ_{dn}.

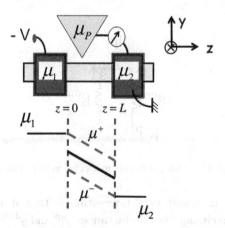

Fig. 12.8 Spatial profile of QFL's μ^+ and μ^- and electrochemical potential μ across a diffusive channel. Also shown is a voltage probe used to measure the local potential.

But why is it more exciting to have separate quasi-Fermi levels for up and down spins than for right and left-moving electrons? Because there is no simple way to measure the latter, but the progress in the last two decades has shown that *magnetic contacts can be used to measure the former.* Let me explain.

12.2.2 *Measuring the spin voltage*

Consider the factors that determine the potential μ_P measured by a probe like the one shown in Fig. 12.8. We can use a circuit representation (Fig. 10.4) similar to the one we introduced for weakly coupled non-invasive probes in Chapter 10 (see Eq. (10.4)).

There we saw that an external probe communicates with right moving and left moving states through conductances g^+ and g^-. Similarly we could model an external probe as communicating with the two spin channels through conductances g^{up} and g^{dn} as shown in Fig. 12.9, so that setting the probe current I_P equal to zero, we have from simple circuit theory

$$I_P = 0 = g^{up}(\mu^{up} - \mu_P) + g^{dn}(\mu^{dn} - \mu_P)$$

$$\rightarrow \mu_P = \frac{g^{up}\mu^{up} + g^{dn}\mu^{dn}}{g^{up} + g^{dn}} \tag{12.17}$$

very similar to Eq. (10.4) with g^{up} and g^{dn} in place of g^+ and g^-.

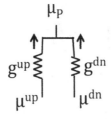

Fig. 12.9   Simple circuit model for voltage probe.

What makes this result very interesting is that it is possible to use magnetic probes to change the conductances g^{up} and g^{dn} simply by rotating their magnetization as established through the tremendous progress in the field of spintronics in the last twenty five years.

Defining the average potential μ and the spin potential μ_S as

$$\mu = \frac{\mu^{up} + \mu^{dn}}{2} \tag{12.18a}$$

$$\mu_S = \frac{\mu^{up} - \mu^{dn}}{2} \tag{12.18b}$$

we can rewrite Eq. (12.17) with a little algebra as

$$\mu_P = \mu + P\frac{\mu_S}{2} \tag{12.19a}$$

$$P \equiv \frac{g^{up} - g^{dn}}{g^{up} + g^{dn}} \tag{12.19b}$$

where P denotes the probe polarization. A non-magnetic probe has equal conductances g^{up} and g^{dn} for both spins making the polarization P equal to zero. But magnets have unequal density of states at the Fermi energy for up and down spins resulting in unequal conductances g^{up} and g^{dn} and hence a non-zero P which could be positive or negative depending on the magnet.

If we reverse the magnetization of the magnet, the sign of P will reverse (either negative to positive, or positive to negative) since the role of up and down are reversed. Using this fact we can write from Eq. (12.19)

$$\mu_P(+\mathbf{m}) - \mu_P(-\mathbf{m}) = 2P\mu_S \tag{12.20}$$

which affords a straightforward approach for measuring the spin potential μ_S, simply by looking at the change in the probe potential on reversing its magnetization.

12.2.3 *Spin-momentum locking*

Let us now get back to our earlier discussion about a special class of materials called topological insulators in which the QFL's for right- and left-moving states translate into those for up and down states which can then be measured with a magnetic probe as we just discussed. However, this translation from μ^+ and μ^- to μ^{up} and μ^{dn} occurs more generally in a large class of materials with strong "spin-orbit (SO) coupling" (we will discuss this more in Part B) which exhibit surface states that have unequal numbers M and N of up-spin and down spin modes propagating to the right (Fig. 12.10).

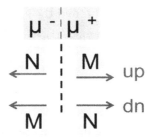

Fig. 12.10 Surface states in materials with high spin-orbit coupling have equal number of modes M for right moving upspins and leftmoving downspins, but a different number of modes N for left moving upspins and right moving downspins.

Note that the situation depicted in Fig. 12.10 is different from ordinary materials as well as from magnetic materials. Ordinarily the number of modes is the same for left-moving and right-moving states for both up and down spins:

$$M(up^+) = M(dn^+) = M(up^-) = M(dn^-) = M.$$

Magnetic materials on the other hand have different numbers of modes for upspins and downspins:

$$M(up^+) = M(up^-) = M; \quad M(dn^+) = M(dn^-) = N.$$

What we are discussing here is different (Fig. 12.10)

$$M(up^+) = M(dn^-) = M; \quad M(dn^+) = M(up^-) = N.$$

This situation arises in non-magnetic materials with high spin-orbit coupling which we will discuss in Part B. For the moment let us accept the picture shown in Fig. 12.10 and note that in these materials, the upspin and downspin potentials μ^{up} and μ^{dn} represent different averages of μ^+ and μ^-

$$\mu^{up} = \frac{M\mu^+ + N\mu^-}{M + N} \quad \text{and} \quad \mu^{dn} = \frac{N\mu^+ + M\mu^-}{M + N} \tag{12.21}$$

which yields a spin potential (see Eq. (12.18))

$$\mu_S = \frac{\mu^{up} - \mu^{dn}}{2} = p\frac{\mu^+ - \mu^-}{2} \tag{12.22}$$

where we have defined the channel polarization as

$$p \sim \frac{M - N}{M + N}. \tag{12.23}$$

In Eq. (12.23) we are not using the equality sign since we have glossed over a "little" detail involving the fact that right moving electrons travel in different directions along the surface, so that their spins also have an angular distribution, which on averaging gives rise to a numerical factor.

Making use of Eqs. (8.8) and (4.12) we can write

$$I = G_B \frac{\mu^+ - \mu^-}{q} \tag{12.24}$$

we can rewrite the spin potential from Eq. (12.22) in terms of the current I:

$$\mu_S = \frac{q}{2G_B} p I. \tag{12.25}$$

Note that the channel polarization "p" appearing in Eq. (12.25) is a channel property that determines the intrinsic spin potential appearing in the channel. It is completely different from the probe polarization "P" defined in Eq. (12.19) which is a magnet property that comes into the picture only when we use a magnetic probe to measure the intrinsic spin potential μ_S induced in the channel by the flow of current (I).

This is a remarkable result that shows a new way of generating spin potentials. The spin valves discussed in Section 12.2.1 generated spin potentials through the spin-dependent interface resistance of magnetic contacts. By contrast Eq. (12.25) tells us that a spin voltage can be generated in channels with spin-momentum locking simply by the flow of current without the need for magnetic contacts, arising from the difference between M and N.

This is our view of the Rashba-Edelstein (RE) effect which has been observed in a wide variety of materials like topological insulators and narrow gap semiconductors. Similar effects are also observed in heavy metals where it is called the spin Hall effect (SHE) and is often associated with bulk scattering mechanisms, but there is some evidence that it could also involve the surface mechanism described here. We will not discuss this current-induced spin potential any further since our understanding is still evolving.

We mention it here simply because it connects spin voltages to the notion of quasi-Fermi levels that we have been discussing in the last few chapters and also gives the reader a feeling for the amazing progress in spintronics that has made it possible to control and measure spin potentials.

Note that in this chapter we are using a semiclassical picture that regards up and down spins simply as two types of electrons, like "red" and "blue" electrons. This picture allows us to understand many spin-related phenomena, but not all. Many phenomena involve additional subtleties that require the quantum picture and hence can only be discussed in depth in Part B.

12.3   Concluding Remarks

Throughout this book we have discussed how the contacts in an ordinary device drive drainbound and sourcebound states out of equilibrium faster than backscattering processes can restore equilibrium. The primary message I hope to convey in this part is that QFL's are quite real and can be generated and measured through the use of "smart contacts". We illustrate this with several examples.

In p-n junctions, contacts drive the two bands out of equilibrium, faster than R-G processes can restore equilibrium. In spin valve devices magnetic contacts drive upspin and downspin states out of equilibrium faster than spin-flip processes can restore equilibrium. *In either case there are groups of states A, B etc that are driven out of equilibrium by smart contacts that can discriminate between them.*

On the other hand in materials with high spin-orbit coupling, a current injected through ordinary contacts generates a spin potential due to the phenomenon of "spin-momentum locking" leading to unequal values of M and N (Fig. 12.10). But to detect the spin potential we need a smart contact. This observation is related to the Rashba-Edelstein (RE) and perhaps the spin Hall effect (SHE) as discussed earlier.

Alternatively we could reverse the voltage and current terminals and invoke reciprocity (Section 10.3.3) to argue that a current injected through a smart contact will generate a voltage at the ordinary contacts. This is related to the inverse Rashba-Edelstein (IRE) and perhaps the inverse spin Hall effect (ISHE).

More and more of such examples can be expected in the coming years, as we learn to control current flow not just with gate electrodes that control the electrostatic potential, but with subtle contacting schemes that engineer the electrochemical potential(s). Many believe that nature does just that in designing many biological "devices", but that is a different story.

In the context of man-made devices there are many possibilities. Perhaps we will figure out how to contact s-orbitals differently from p-orbitals, or one valley differently from another valley, leading to fundamentally different devices. But this requires a basic change in approach.

Traditionally, the work of device design has been divided neatly between three groups of specialists: physicists and material scientists who innovate new materials using atomistic theory, device engineers who worry about contacts and related issues using macroscopic theory and circuit designers who interconnect devices to perform useful functions.

Future devices that seek to function effectively may well require an approach that integrates materials, contacts and even circuits at the atomistic level. Perhaps then we will be able to create devices that rival the marvels of nature like photosynthesis.

PART 4

Heat and Electricity

Chapter 13

Thermoelectricity

13.1 Introduction

Conductance measurements ordinarily do not tell us anything about the nature of the conduction process inside the conductor. If we connect the terminals of a battery across any conductor, electron current flows out of the negative terminal back to its positive terminal. Since this is true of all conductors, it clearly does not tell us anything about the conductor itself.

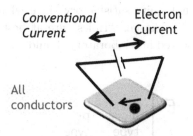

On the other hand, thermoelectricity, that is, electricity driven by a temperature difference, is an example of an effect that does. A very simple experiment is to look at the current between a hot probe and a cold probe (Fig. 13.1). For an n-type conductor (see Fig. 6.1) the direction of the external current will be consistent with what we expect if electrons travel from the hot to the cold probe inside the conductor, but for a p-type conductor (see Fig. 6.2) the direction is reversed, consistent with electrons traveling from the cold to the hot probe. Why?

It is often said that p-type conductors show the opposite effect because the carriers have the opposite sign. As we discussed in Chapter 6, p-type conductors involve the flow of electrons near the top of a band of energies and it is convenient to keep track of the empty states above μ rather than the filled states below μ. These empty states are called "holes" and since they represent the absence of an electron, they behave like positively charged entities.

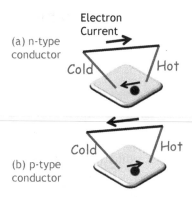

Fig. 13.1 Thermoelectric currents driven by temperature differences flow in opposite directions for n- and p-type conductors.

However, this is not quite satisfactory since what moves is really an electron with a negative charge. "Holes" are at best a conceptual convenience and effects observed in a laboratory should not depend on subjective conveniences.

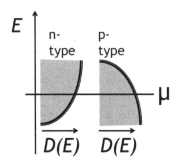

Fig. 13.2 In n-type conductors the electrochemical potential is located near the bottom of a band of energies, while in p-type conductors it is located near the top. In n-conductors $D(E)$ increases with increasing E, while in p-conductors it decreases with increasing E.

As we will see in this chapter the difference between n- and p-conductors requires no new principles or assumptions beyond what we have already discussed, namely that the current is driven by the difference between f_1 and f_2. The essential difference between n- and p-conductors is that while one has a density of states $D(E)$ that increases with energy E, the other has a $D(E)$ decreasing with E.

Earlier in Chapter 11 we discussed the Hall effect which too changes sign for n-type and p-type conductors and this too is commonly blamed on negative and positive charges. The Hall effect, however, has a totally different origin related to the negative mass ($m = p/v$) associated with $E(p)$ relations in p-conductors that point downwards. By contrast the thermoelectric effect does not require a conductor to even have a $E(p)$ relation. Even small molecules show sensible thermoelectric effects (Baheti *et al.* 2008).

The basic idea is easy to see starting from our old expression for the current obtained in Chapter 3:

$$I = \frac{1}{q} \int_{-\infty}^{+\infty} dE\, G(E)(f_1(E) - f_2(E)) \quad \text{(same as Eq. (3.3))}. \tag{13.1}$$

So far the difference in f_1 and f_2 has been driven by difference in electrochemical potentials μ_1 and μ_2. But it could just as well be driven by a temperature difference, since in general

$$f_1(E) = \frac{1}{\exp\left(\dfrac{E - \mu_1}{kT_1}\right) + 1} \tag{13.2}$$

and

$$f_2(E) = \frac{1}{\exp\left(\dfrac{E - \mu_2}{kT_2}\right) + 1}. \tag{13.3}$$

But why would such a current reverse directions for an n-type and a p-type conductor? To see this, consider two contacts with the same electrochemical potential μ, but with different temperatures as shown in Fig. 13.3.

The key point is that the difference between $f_1(E)$ and $f_2(E)$ has a different sign for energies E greater than μ and for energies less than μ (see Fig. 13.3). In an n-type channel, the conductance $G(E)$ is an increasing function of energy, so that the net current is dominated by states with energy $E > \mu$ and thus flows from 1 to 2, that is from hot to cold (Fig. 13.4). But in a p-type channel it is the opposite. The conductance $G(E)$ is a decreasing function of energy, so that the net current is dominated by states with energy $E < \mu$ and thus flows from 2 to 1, that is from cold to hot.

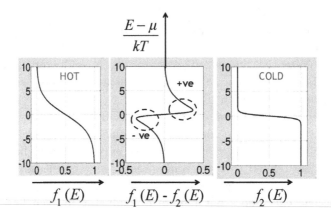

Fig. 13.3   Two contacts with the same μ, but different temperatures: $f_1 - f_2$ is positive for $E > \mu$, and negative for $E < \mu$.

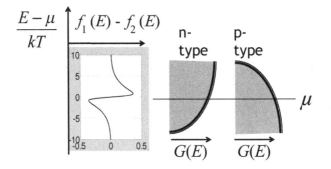

Fig. 13.4   For n-type channels, the current for $E > \mu$ dominates that for $E < \mu$, while for p-type channels the current for $E < \mu$ dominates that for $E > \mu$. Consequently, electrons flow from hot to cold across an n-type channel, but from cold to hot in a p-type channel.

13.2   Seebeck Coefficient

Related video lecture available at course website, Unit 4: L4.2.

We can use Eq. (13.1) directly to calculate currents without making any approximations. But it is often convenient to use a Taylor series expansion like we did earlier (Eq. (2.10)) to obtain results that are reasonably accurate

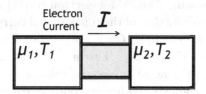

for low "bias". We could write approximately from Eq. (13.1)

$$I = G_0(V_1 - V_2) + G_S(T_1 - T_2) \tag{13.4}$$

where we have defined V_1 and V_2 as μ_1/q and μ_2/q. The conductance is given by

$$G_0 = \int_{-\infty}^{+\infty} dE\, G(E) \left(\frac{\partial f_0}{\partial \mu} \right)$$

$$= \int_{-\infty}^{+\infty} dE \left(-\frac{\partial f_0}{\partial E} \right) G(E) \tag{13.5}$$

as we have seen before in Section 2.4. The new coefficient G_S that we have introduced is given by

$$G_S = \frac{1}{q} \int_{-\infty}^{+\infty} dE\, G(E) \left(\frac{\partial f_0}{\partial T} \right)$$

$$= \int_{-\infty}^{+\infty} dE \left(-\frac{\partial f_0}{\partial E} \right) \frac{E - \mu_0}{qT} G(E). \tag{13.6}$$

This last step, relating the derivatives with respect to T and with respect to E, requires a little algebra (see Appendix A).

Equation (13.6) expresses mathematically the basic point we just discussed. Energies E greater and less than μ_0, contribute with opposite signs to the thermoelectric coefficient, G_S. It is clear that if we wanted to design a material with the best Seebeck coefficient, S we would try to choose a material with all its density of states on one side of μ_0 since anything on the other side contributes with an opposite sign and brings it down. We can visualize Eq. (13.4) as shown in Fig. 13.5, where the short circuit current is given by

$$I_{SC} = G_S(T_1 - T_2). \tag{13.7}$$

Experimentally what is often measured is the open circuit voltage

$$V_1 - V_2 = V_{OC} = -\frac{I_{SC}}{G_0} = -\frac{G_S}{G_0}(T_1 - T_2). \tag{13.8}$$

Note that we are using I and V for electron current and electron voltage μ/q whose sign is opposite that of the conventional current and voltage. For n-type conductors, for example, G_S is positive, so that Eq. (13.8) tells us that V_{OC} is negative if $T_1 > T_2$. This means that the contact with the higher temperature has a negative electron voltage (see Section 3.2.2 where our convention is explained) and hence a positive conventional voltage. By convention this is defined as a negative Seebeck coefficient.

$$S \equiv \frac{V_{OC}}{T_1 - T_2} = -\frac{G_S}{G_0} \tag{13.9}$$

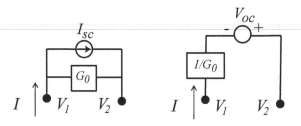

Fig. 13.5 Circuit representations of Eq. (13.4).

13.3 Thermoelectric Figures of Merit

The practical importance of thermoelectric effects arise from the possibility of converting waste heat into electricity and from this point of view the important figure of merit is the amount of power that could be generated from a given $T_1 - T_2$. What load resistor R_L will maximize the power delivered to it (Fig. 13.6)? A standard theorem in circuit theory says (this is not too hard to prove for yourself) that the answer is a "matched load" for which R_L equal to $1/G_0$:

$$P_{max} = \frac{V_{OC}{}^2 G_0}{4} = S^2 G_0 \frac{(T_1 - T_2)^2}{4}. \tag{13.10}$$

The quantity $S^2 G_0$ is known as the power factor and is one of the standard figures of merit for thermoelectric materials.

However, there is a second figure of merit that is more commonly used. To see where this comes from, we first note that when the contacts are at different temperatures, we expect a constant flow of heat through the conductor due to its *heat conductance* G_K

$$G_K(T_1 - T_2)$$

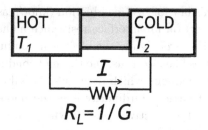

Fig. 13.6 A thermoelectric generator can convert a temperature difference into an electrical output.

which has to be supplied by the source that maintains the temperature difference. Actually this is not quite right, it only gives the heat flow under open circuit conditions and ignores a component that depends on I. But this is good enough for our purpose which is simply to provide an intuitive feeling for where the standard thermoelectric figure of merit comes from.

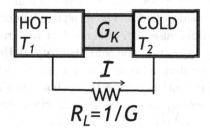

The ratio of the maximum generated power to the power that is supplied by the external source is a good measure of the efficiency of the thermoelectric material in converting heat to electricity and can be written as

$$\frac{P_{max}}{G_K(T_1 - T_2)} = \underbrace{\frac{S^2 G_0 T}{G_K}}_{\equiv ZT} \frac{T_1 - T_2}{4T} \tag{13.11}$$

where T is the average temperature $(T_1 + T_2)/2$. The standard figure of merit for thermoelectric materials, called its ZT *product*, is proportional to the ratio of $S^2 G_0$ to G_K:

$$ZT \equiv \frac{S^2 G_0 T}{G_K} = \frac{S^2 \sigma_0 T}{\kappa} \tag{13.12}$$

where κ is the thermal conductivity related to the thermal conductance G_K by same geometric factor A/L connecting the corresponding electrical quantities G_0 and σ_0.

Indeed the Ohm's law for heat conduction (known as Fourier's law) also needs the same correction for interface resistance namely the replacement of L with $L + \lambda$. However, while the electrical conductivity arises solely from charged particles like electrons, the thermal conductivity also includes a contribution from phonons which describes the vibrations of the atoms comprising the solid lattice. Ordinarily it is the phonon component that dominates the thermal conductivity and we will discuss it briefly in the next chapter. For the moment let us talk about the heat carried by electrons, something we have not discussed so far at all.

13.4   Heat Current

Related video lecture available at course website, Unit 4: L4.3.

We have discussed the thermoelectric currents in a material with any arbitrary conductance function $G(E)$. The nice thing about the elastic resistor is that channels at different energies all conduct in parallel, so that we can think of one energy at a time and add them up at the end. Consider a small energy range located between E and $E + dE$, either above or below the electrochemical potentials μ_1 and μ_2 as shown in Fig. 13.7. As we discussed in the introduction, these two channels will make contributions with opposite signs to the Seebeck effect. Now, it has long been known that the Seebeck effect is associated with a Peltier effect. How can we understand this connection?

Earlier in Chapter 3 we saw that for an elastic resistor the associated Joule heat I^2R is dissipated in the contacts (see Fig. 3.2). But if we consider the n-type or p-type channels in Fig. 13.7, it is apparent that unlike Fig. 3.2, both contacts do not get heated.

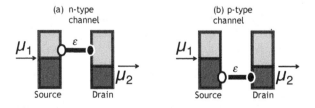

Fig. 13.7   A one-level elastic resistor having just one level with $E = \epsilon$, (a) above or (b) below the electrochemical potentials $\mu_{1,2}$.

Figure 13.8 is essentially the same as Fig. 3.2 except that we have shown the heat absorbed from the surroundings rather than the heat dissipated. For n-type conductors the heat absorbed is positive at the source, negative at the drain, indicating that the source is cooled and the drain is heated. For p-type conductors it is exactly the opposite. This is the essence of the Peltier effect that forms the basis for practical thermoelectric refrigerators. Note that the sign of the Peltier coefficient like that of the Seebeck coefficient is related to the sign of $E - \mu$ and not the sign of q.

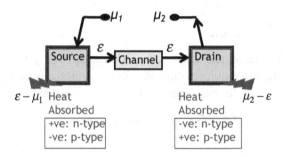

Fig. 13.8 Same as Fig. 3.2 but showing the heat absorbed (rather than dissipated) at each contact. For n-type conductors the heat absorbed is positive at the source, negative at the drain showing that the electrons COOL the source and HEAT the drain. For p-type conductors it is exactly the opposite.

To write the heat current carried by electrons, we can simply extend what we wrote for the ordinary current earlier:

$$I = \frac{1}{q} \int_{-\infty}^{+\infty} dE \, G(E)(f_1(E) - f_2(E)) \text{ (same as Eq. (3.3))}.$$

Noting that an electron with energy E carrying a charge $-q$ also extracts an energy $E - \mu_1$ from the source and dumps an energy $E - \mu_2$ in the drain, we can write the heat currents I_{Q1} and I_{Q2} *extracted* from the source and drain respectively as

$$I_{Q1} = \frac{1}{q} \int_{-\infty}^{+\infty} dE \, \frac{E - \mu_1}{q} G(E)(f_1(E) - f_2(E)) \qquad (13.13)$$

$$I_{Q2} = \frac{1}{q} \int_{-\infty}^{+\infty} dE \, \frac{\mu_2 - E}{q} G(E)(f_1(E) - f_2(E)). \qquad (13.14)$$

The energy extracted from the external source per unit time is given by

$$I_E = VI = \frac{\mu_1 - \mu_2}{q} I. \qquad (13.15)$$

Making use of the current equation Eq. (3.3) we can rewrite I_E in the form

$$I_E = \frac{1}{q} \int_{-\infty}^{+\infty} dE \, \frac{\mu_1 - \mu_2}{q} \, G(E)(f_1(E) - f_2(E))$$

which can be combined with the equations for I_{Q1} and I_{Q2} above to show that the sum of all three energy currents is zero

$$I_{Q1} + I_{Q2} + I_E = 0$$

as we would expect due to overall energy conservation.

13.4.1 *Linear response*

Just as we linearized the current equation (Eq. (3.3)) to obtain an expression for the current in terms of voltage and temperature differences (Eq. (13.4)), we can linearize the heat current equation to obtain

$$I_Q = G_P(V_1 - V_2) + G_Q(T_1 - T_2) \tag{13.16}$$

$$\text{where} \quad G_P = \int_{-\infty}^{+\infty} dE \left(-\frac{\partial f_0}{\partial E} \right) \frac{E - \mu_0}{q} \, G(E) \tag{13.17a}$$

$$G_Q = \int_{-\infty}^{+\infty} dE \left(-\frac{\partial f_0}{\partial E} \right) \frac{(E - \mu_0)^2}{q^2 T} \, G(E). \tag{13.17b}$$

These are the standard expressions for the thermoelectric coefficients due to electrons which are usually obtained from the Boltzmann equation.

I should mention that the quantity G_Q we have obtained is not the thermal conductance G_K that is normally used in the ZT expression cited earlier (Eq. (13.12)). One reason is what we have stated earlier, namely that G_K also has a phonon component that we have not yet discussed. But there is another totally different reason. The quantity G_K is defined as the heat conductance under electrical open circuit conditions ($I = 0$):

$$G_K = \left(\frac{\partial I_Q}{\partial (T_1 - T_2)} \right)_{I=0}$$

while it can be seen from Eq. (13.16) that G_Q is the heat conductance under electrical short circuit conditions ($V = 0$):

$$G_Q = \left(\frac{\partial I_Q}{\partial (T_1 - T_2)} \right)_{V_1 = V_2}.$$

However, we can rewrite Eqs. (13.4) and (13.16) in a form that gives us the open circuit coefficients (as noted in Fig. 3.3, V and I represent the electron voltage μ/q and the electron current, which are opposite in sign to the conventional voltage and current)

$$(V_1 - V_2) = \frac{1}{G_0} \times I + \overbrace{\left(-\frac{G_S}{G_0}\right)}^{s,\ \text{Seebeck}} \times (T_1 - T_2) \quad (13.18a)$$

$$I_Q = \underbrace{\frac{G_P}{G_0}}_{\text{Peltier, } \Pi} \times I - \underbrace{\left(G_Q - \frac{G_P G_S}{G_0}\right)}_{\text{Heat Conductance, } G_K} \times (T_1 - T_2). \quad (13.18b)$$

We have indicated the coefficients that are normally measured experimentally and are named after the experimentalists who discovered them. Eqs. (13.4) and (13.16), on the other hand, come more naturally in theoretical models because of our Taylor series expansion and it is important to be aware of the difference. Incidentally, using the expressions in Eqs. (13.6) and (13.17), we can see that the Peltier and Seebeck coefficients in Eq. (13.18) obey the Kelvin relation

$$\Pi = TS \quad (13.19)$$

which is a special case of the fundamental Onsager relations that the linear coefficients are required to obey (Section 10.3.4).

13.5 The Delta Function Thermoelectric

Related video lecture available at course website, Unit 4: L4.4.

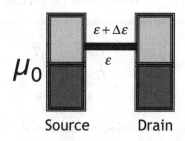

It is instructive to look at a so-called "delta function" thermoelectric, which is a hypothetical material with a narrow conductance function located at energy ϵ with a width $\Delta\epsilon$ that is much less than kT. It is straightforward to obtain the thermoelectric coefficients of this delta function thermoelectric formally starting from the general relations we have obtained in this chapter, reproduced below for convenience:

$$G_0 = \int_{-\infty}^{+\infty} dE \left(-\frac{\partial f_0}{\partial E} \right) G(E) \text{ (same as Eq. (13.5))}$$

$$G_S = \int_{-\infty}^{+\infty} dE \left(-\frac{\partial f_0}{\partial E} \right) \frac{E - \mu_0}{qT} G(E) \text{ (same as Eq. (13.6))}$$

$$G_P = \int_{-\infty}^{+\infty} dE \left(-\frac{\partial f_0}{\partial E} \right) \frac{E - \mu_0}{q} G(E)$$

$$G_Q = \int_{-\infty}^{+\infty} dE \left(-\frac{\partial f_0}{\partial E} \right) \frac{(E - \mu_0)^2}{q^2 T} G(E) \text{ (same as Eq. (13.17))}.$$

We argue that factors like $E - \mu_0$ can be pulled out of the integrals assuming they are almost constant over the very narrow energy range where $G(E)$ is non-zero. This gives

$$G_0 = G(\varepsilon) \Delta\varepsilon \left(-\frac{\partial f_0}{\partial E} \right)_{E=\varepsilon} \tag{13.20a}$$

$$G_S = \left(\frac{\varepsilon - \mu_0}{qT} \right) G_0 \tag{13.20b}$$

$$G_P = \left(\frac{\varepsilon - \mu_0}{q} \right) G_0 \tag{13.20c}$$

$$G_Q = \left(\frac{(\varepsilon - \mu_0)^2}{q^2 T} \right) G_0. \tag{13.20d}$$

From Eqs. (13.18) and (13.20) we obtain the coefficients for the *delta function thermoelectric*:

$$S = -\frac{G_S}{G_0} = -\frac{\varepsilon - \mu_0}{qT} \tag{13.21a}$$

$$\Pi = -\frac{G_P}{G_0} = -\frac{\varepsilon - \mu_0}{q} \tag{13.21b}$$

$$G_K = G_Q - \frac{G_P G_S}{G_0} = 0. \tag{13.21c}$$

Let us now see how we can understand these results from intuitive arguments without any formal calculations. The Seebeck coefficient in Eq. (13.21a) is the open circuit voltage required to maintain zero current. Since the channel conducts only at a single energy $E = \varepsilon$, in order for no current to flow, the Fermi functions at this energy must be equal:

$$f_1(\varepsilon) = f_2(\varepsilon) \rightarrow \frac{\varepsilon - \mu_1}{kT_1} = \frac{\varepsilon - \mu_2}{kT_2}.$$

Hence

$$\frac{\varepsilon - \mu_1}{kT_1} = \frac{\varepsilon - \mu_2}{kT_2} = \frac{(\varepsilon - \mu_1) - (\varepsilon - \mu_2)}{k(T_1 - T_2)} = -\frac{\mu_1 - \mu_2}{k(T_1 - T_2)}.$$

Noting that the Seebeck coefficient is defined as

$$S \equiv \frac{(\mu_1 - \mu_2)/q}{(T_1 - T_2)} \text{ (with } I = 0)$$

we obtain

$$S = -\frac{\varepsilon - \mu_1}{qT_1} = -\frac{\varepsilon - \mu_2}{qT_2} \approx -\frac{(\varepsilon - \mu_0)}{qT}$$

in agreement with the result in Eq. (13.21a).

The expression in Eq. (13.21b) for the Peltier coefficient too can be understood in simple terms by arguing that every electron carries a charge $-q$ and a heat $\varepsilon - \mu_0$, and hence the ratio of the heat current to the charge current must be $(\varepsilon - \mu_0)/(-q)$.

That brings us to the zero current heat conductance in Eq. (13.21c) which tells us that the heat current is zero under open circuit conditions, that is when the regular charge current is zero. This seems quite reasonable. After all if there is no electrical current, how can there be a heat current? But if this were the whole story, then no thermoelectric would have any heat conductance, and not just delta function thermoelectrics.

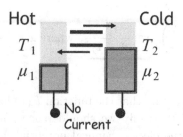

The full story can be understood by considering a two-channel thermo-electric with a temperature difference. Under open circuit conditions, there is a voltage between the two contacts with $\mu_1 < \mu_2$. Although the total current is zero, the individual currents in each level are non-zero. They are equal and opposite, thereby canceling each other out. But the corresponding energy currents do not cancel, since the channel with higher energy carries more energy. Zero charge current thus does not guarantee zero heat current, except for a delta function thermoelectric with its sharply peaked $G(E)$.

Since the delta function thermoelectric has zero heat conductance, the ZT product (see Eq. (13.12)) should be very large and it would seem that is what an ideal thermoelectric should look like. However, as we mentioned earlier, even if the electronic heat conductance were zero, we would still have the phonon contribution which would prevent the ZT-product from getting too large. We will talk briefly about this aspect in the next chapter.

13.5.1 *Optimizing power factor*

Let us end this chapter by discussing what factors might maximize the power factor S^2G_0 (see Eq. (13.10)) for a thermoelectric. If getting the highest Seebeck coefficient S were our sole objective then it is apparent from Eq. (13.21a) that we should choose our energy ε as far from μ_0 as possible. But that would make the conductance G_0 from Eq. (13.20a) unacceptably low, because the factor $-(\partial f_0/\partial E)$ dies out quickly as the energy E moves away from μ_0.

From Eqs. (13.20) and (13.21a) we have

$$S^2G_0 = G(\varepsilon)\Delta\varepsilon\left(\frac{\varepsilon - \mu_0}{qT}\right)^2\left(-\frac{\partial f_0}{\partial E}\right)_{E=\varepsilon}$$

$$= G(\varepsilon)\frac{\Delta\varepsilon}{kT}\left(\frac{k}{q}\right)^2\underbrace{x^2\frac{e^x}{(e^x + 1)^2}}_{\equiv F(x)}, \quad \text{where } x \equiv \frac{\varepsilon - \mu_0}{kT}. \tag{13.22}$$

It is apparent from Fig. 13.9 that the function $F(x)$ has a maximum around $x \sim 2$, suggesting that ideally we should place our level approximately $2kT$ above or below the electrochemical potential μ_0. The corresponding

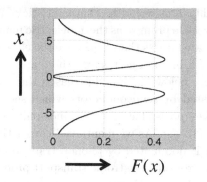

Fig. 13.9   Plot of $F(x) \equiv \frac{x^2 e^x}{(e^x+1)^2}$.

Seebeck coefficient and power factor are approximately given by

$$S \approx 2\frac{k}{q} \qquad (13.23\text{a})$$

$$S^2 G_0 \approx 0.5\left(\frac{k}{q}\right)^2 G(\varepsilon)\frac{\Delta\varepsilon}{kT}. \qquad (13.23\text{b})$$

The best thermoelectrics typically have Seebeck coefficients that are not too far from the $2(k/q) = 170\ \mu\text{V/K}$ expected from Eq. (13.23a). They are usually designed to place μ_0 a little below the bottom of the band so that the product of $G(E)$ and $-(\partial f_0/\partial E)$ looks like a "delta function" around ϵ a little above the bottom of the band as shown in the sketch on the left.

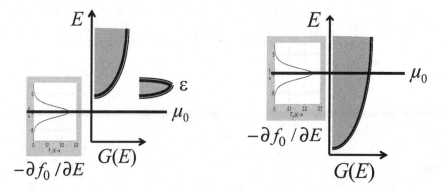

The problem is that the corresponding values of conductance are not as large as they could possibly be if μ_0 were located higher up in the band as sketched on the right. This would be characteristic of metals.

But metals show little promise as thermoelectric materials, because their Seebeck coefficients are far less than k/q, since the electrochemical potential lies in the middle of a band of states and there are nearly as many states above μ_0 as there are below μ_0. For this reason the field of thermoelectric materials is dominated by semiconductors which show the highest power factors. However, the power factor determines only the numerator of the ZT product in Eq. (13.12). As we mentioned earlier the heat conductance in the denominator is dominated by phonon transport involving a physics that is very different from the electronic transport properties that this book is largely about. In the next chapter we will digress briefly to talk about phonon transport.

Chapter 14

Phonon Transport

14.1 Introduction

We have seen earlier that the electrical conductivity is given by (Eq. (6.22))

$$\sigma = \frac{q^2}{h}\left(\frac{M\lambda}{A}\right) \tag{14.1}$$

where the number of channels per unit area M/A and the mean free path λ are evaluated in an energy window $\sim$ a few kT around μ_0. The degeneracy factor g (Section 6.4.1) due to spins and valleys is assumed to be included in M.

In this chapter we will extend the transport theory for electrons to handle something totally different, namely phonons and obtain a similar expression for the thermal conductivity due to phonons

$$\kappa_{ph} = \frac{\pi^2}{3}\frac{k^2 T}{h}\left(\frac{M\lambda}{A}\right)_{ph} \tag{14.2}$$

where the number of channels per unit area M/A and the mean free path λ are evaluated in a frequency window $\hbar\omega \sim$ a few kT. There is a degeneracy factor of $g = 3$ due to three polarizations that is assumed to be included in M.

Our purpose in this chapter is two-fold. The first is to provide an interesting perspective in the hunt for high-ZT thermoelectrics. We have seen in Chapter 13 that with careful design it is possible to achieve a Seebeck coefficient $\sim 2k/q$ while maximizing the numerator in Eq. (13.12). We can write

$$ZT \approx 4\frac{k^2 T}{q^2}\left(\frac{\sigma}{\kappa + \kappa_{ph}}\right) \approx 4\frac{k^2 T}{q^2}\frac{\sigma}{\kappa_{ph}} \tag{14.3}$$

if we assume that the thermoelectric has been designed to provide a Seebeck coefficient $S \sim 2k/q$ and the heat conductivity is dominated by phonons. Using Eq. (14.3) with Eqs. (14.1) and (14.2) we have

$$ZT \approx \frac{M\lambda/A}{(M\lambda/A)_{ph}} \qquad (14.4)$$

where we have dropped a factor of $12/\pi^2 \sim 1$ since it is just an approximate number anyway.

This is an interesting expression suggesting that once a material has been optimized to provide a respectable Seebeck coefficient (S), the ZT product we obtain simply reflects the ratio of $M\lambda/A$ for electrons and phonons. As we discussed at the end of the last chapter, the process of achieving a high S usually puts us in a regime with a low M/A for electrons. M/A for phonons on the other hand is often much higher ~ 1 nm^{-2} at room temperature, so that the ratio of M/A's in Eq. (14.4) is ~ 0.1 or less. But electrons tend to have a longer mean free path, resulting in a $ZT \sim 1$ for the best thermoelectrics.

The most promising approach for improving ZT at this time seems to be to try to suppress the mean free path for phonons without hurting the electrons (the so-called "electron crystal, phonon glass"). The highest ZT was about 1 for a long time and has recently increased to 3. Experts say that an increase of ZT to 4 to 10 would have a major impact on its practical applications and researchers hope that nanostructured materials might enable this increase.

Whether they are right, only future experiments can tell, but it is clearly of interest to understand the principles that govern ZT in nanoscale materials and we hope this chapter will contribute to this understanding. But my real objective is to demonstrate the power of the elastic resistor approach that allows us not only to obtain linear transport coefficients for electrons easily, but also extend the results to a totally different entity (the phonons) with relative ease.

14.2 Phonon Heat Current

As we mentioned earlier the thermal conductance of solids has a significant phonon component in addition to the electronic component we just talked about. I will not go into this in any depth. My purpose is simply to show how easily our elastic transport model is extended to something totally different.

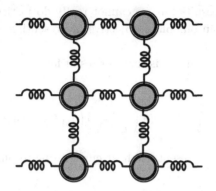

The atoms comprising the solid lattice are often pictured as an array of masses connected by springs as sketched above. The vibrational frequencies of such a system are described by a dispersion relation $\omega(\beta)$ analogous to the $E(k)$ relation that describes electron waves, with β playing the role of k, and $\hbar\omega$ playing the role of E. The key difference with electrons is that unlike electrons, there is no exclusion principle. Millions of phonons can be packed into a single channel creating a sound wave that we can even hear, if the frequency is low enough.

One consequence of this lack of an exclusion principle is that the equilibrium distribution of phonons is given by a Bose function

$$n(\omega) \equiv \frac{1}{\exp\left(\dfrac{\hbar\omega}{kT}\right) - 1} \tag{14.5}$$

instead of the Fermi function for electrons introduced in Chapter 2

$$f(E) \equiv \frac{1}{\exp\left(\dfrac{E - \mu}{kT}\right) + 1} \quad \text{(same as Eq. (2.2))}.$$

One difference with Eq. (14.5) is just the $+1$ instead of the -1 in the denominator, which restricts $f(E)$ to values between 0 and 1, unlike the $n(\omega)$ in Eq. (14.5). The other difference is the absence of an electrochemical potential μ in Eq. (14.5) which is attributed to the lack of conservation of phonons. Unlike electrons, they can appear and disappear as long as other entities are around to take care of energy conservation.

These results are of course not meant to be obvious, but they represent basic results from equilibrium statistical mechanics that are discussed in standard texts. In Chapter 15 on the Second Law, we will try to say a little more about the basics of equilibrium statistical mechanics. We make use of these equilibrium results but we cannot really do justice to them without a major detour from our main objective of presenting a new approach to non-equilibrium problems.

Anyway, the bottom line is that our result for the charge current carried by electrons

$$I = \frac{q}{h} \int_{-\infty}^{+\infty} dE \left(\frac{M\lambda}{L+\lambda} \right) \left(f_1(E) - f_2(E) \right)$$

can be modified to represent the heat current due to phonons

$$I_Q = \frac{1}{h} \int_{0}^{+\infty} d(\hbar\omega) \left(\frac{M\lambda}{L+\lambda} \right)_{ph} \hbar\omega \left(n_1(\omega) - n_2(\omega) \right) \qquad (14.6)$$

simply by making the replacements $q \to \hbar\omega$, $E \to \hbar\omega$ and the Fermi functions with the Bose functions:

$$n_1(\omega) \equiv \frac{1}{\exp\left(\dfrac{\hbar\omega}{kT_1} \right) - 1}$$

$$n_2(\omega) \equiv \frac{1}{\exp\left(\dfrac{\hbar\omega}{kT_2} \right) - 1}$$

and changing the lower integration limit to zero.

Again we can linearize Eq. (14.6) to write (see Appendix A)

$$I_Q \approx G_K(T_1 - T_2) \qquad (14.7)$$

where the thermal conductance due to phonons can be written as

$$G_K = \frac{k^2 T}{h} \int_{0}^{+\infty} dx \left(\frac{M\lambda}{L+\lambda} \right)_{ph} \frac{x^2 e^x}{(e^x - 1)^2}, \quad x \equiv \frac{\hbar\omega}{kT}. \qquad (14.8)$$

Note that just as the elastic resistor model for electrons ignores effects due to the inelastic scattering between energy channels, this model for phonons ignores effects due to the so-called "anharmonic interactions" that cause phonons to convert from one frequency to another. While ballistic electron devices have been widely studied for nearly two decades, much less is known about ballistic phonon devices.

14.2.1 *Ballistic phonon current*

Before moving on let us take a brief detour to point out that the ballistic conductance due to phonons is well-known though in a slightly different form, similar to the Stefan-Boltzmann law for photons. From Eq. (14.8) we can write the ballistic heat conductance as

$$
\left[G_\kappa \right]_{ballistic} = \frac{k^2 T}{h} \int_0^{+\infty} dx \; M_{ph} \frac{x^2 e^x}{(e^x - 1)^2}. \tag{14.9}
$$

To evaluate this expression we need to evaluate the number of modes which is related to the number of wavelengths that fit into the cross-section, as we discussed for electrons (see Eq. (6.20))

$$
M_{ph} = \frac{\pi A}{(wavelength)^2} \quad \times \quad \underbrace{3}_{\substack{\text{number of} \\ \text{polarizations}}}
$$

but we have a degeneracy factor of 3 for the three allowed polarizations. Noting that for phonons (c_s: acoustic wave velocity)

$$
wavelength = \frac{c_s}{\omega/2\pi}
$$

we have $\quad M_{ph} = \dfrac{3\omega^2 A}{4\pi c_s^2} = \dfrac{3k^2 T^2 A}{4\pi \hbar^2 c_s^2} x^2.$

From Eq. (14.9),

$$
\left[G_K \right]_{ballistic} = \frac{3k^4 T^3}{8\pi^2 \hbar^3 c_s^2} \underbrace{\int_0^{+\infty} dx \frac{x^4 e^x}{(e^x - 1)^2}}_{=4\pi^4/15} = \frac{\pi^2 k^4 T^3}{10 \hbar^3 c_s^2}.
$$

Making use of this expression we can write the ballistic heat current from Eq. (14.7) as

$$
\left[I_Q \right]_{ballistic} = \frac{\pi^2 k^4 T^3}{10 \hbar^3 c_s^2} \Delta T.
$$

However, the ballistic current is usually written in a different form making use of the relation $T^3 \Delta T = \Delta(T^4/4)$:

$$
\left[I_Q \right]_{ballistic} = \frac{\pi^2 k^4}{40 \hbar^3 c_s^2} \Delta(T^4) = \frac{\pi^2 k^4}{40 \hbar^3 c_s^2} (T_1{}^4 - T_2{}^4). \tag{14.10}
$$

The corresponding result for photons is known as the Stefan-Boltzmann relation for which the numerical factor differs by a factor of 2/3 because the number of polarizations is 2 instead of 3. But this is just a detour. Let us get back to diffusive phonon transport.

14.3   Thermal Conductivity

Returning to Eq. (14.8) for the thermal conductance due to phonons, we could define the thermal conductivity

$$\kappa = \frac{k^2 T}{h} \int_0^{+\infty} dx \left(\frac{M\lambda}{A} \right)_{ph} \frac{x^2 e^x}{(e^x - 1)^2} \qquad (14.11)$$

such that $G_\kappa = \dfrac{\kappa A}{\lambda + L}$.

Note the similarity with the electrical conductivity due to electrons:

$$\sigma = \frac{q^2}{h} \int_{-\infty}^{+\infty} dx \left(\frac{M\lambda}{L + \lambda} \right) \frac{e^x}{(e^x + 1)^2}.$$

The function

$$F_T(x) \equiv \frac{e^x}{(e^x + 1)^2}$$

appearing in all electronic transport coefficients is different from the function

$$\frac{3}{\pi^2} \frac{x^2 e^x}{(e^x - 1)^2}$$

appearing in Eq. (14.11) but they are of similar shape as shown. The factor $3/\pi^2$ is needed to make the area under the curve equal to one, as it is for the broadening function $F_T(x)$ for electrons (see Eq. (2.3)).

So we can think of electrical and thermal conductance at least qualitatively in the same way. Just as the electrical conductivity is given by the product

$$\underbrace{\frac{q^2}{h}}_{\sim 40\mu s} \left(\frac{M\lambda}{A} \right)$$

the thermal conductivity is given by

$$\underbrace{\frac{\pi^2}{3} \frac{k^2 T}{h}}_{284 pW/K} \left(\frac{M\lambda}{A} \right)_{ph}.$$

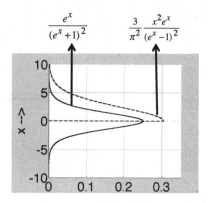

Fig. 14.1 Broadening function for phonons compared to that for electrons, $F_T(x)$. These are the window functions defined by Jeong *et al.* (2011), see Eqs. (7e, f).

The factor $\pi^2/3$ is just the inverse of the $3/\pi^2$ needed to normalize the phonon broadening function. We mentioned at the end of the last chapter that M/A for electrons tends to be rather small for good thermoelectric materials whose electrochemical potential μ lies within a kT of the bottom of the band.

One way to get around this is to use materials where the entire electronic band of energies is a few kT wide, which is unusual. Unfortunately for the phonon band this condition is common, giving an average M/A close to the maximum value. The most popular thermoelectric material Bi_2Te_3 appears to have a phonon bandwidth much less than kT, thus making the average value of M/A for phonons relatively small. The phonon mean free path is also relatively small, helping raise ZT. For example, $M/A \sim 4 \times 10^{17}$ m^{-2}, $\lambda \sim 15$ nm gives $\kappa \sim 2$ W $\cdot$ m$^{-1} \cdot$ K^{-1}, numbers that are approximately representative of Bi_2Te_3.

The possible role of nanostructuring in engineering a better thermoelectric is still a developing story whose ending is not clear. At this time all we can do is to present a different viewpoint that may help us see some new options. And that is what we have tried to do here.

Chapter 15

Second Law

15.1 Introduction

Related video lecture available at course website, Unit 4: L4.5.

Back in Chapter 13, when discussing the heat current carried by electrons we drew a picture (Fig. 13.8) showing the flow of electrons and heat in an elastic resistor consisting of a channel with two contacts (source and drain) with a voltage applied across it (Fig. 15.1a). Figure 15.1b shows a slightly generalized version of the same picture that will be useful for the present discussion: An elastic channel receiving N_1 and N_2 electrons with contacts 1 and 2, held at potentials μ_1 and μ_2 respectively.

Of course both N_1 and N_2 cannot be positive. If N_1 electrons enter the channel from one contact an equal number must leave from the other contact so that

$$N_1 + N_2 = 0. \tag{15.1}$$

For generality, I have also shown an exchange of energy E_0 (but not electrons) with the surroundings at temperature T_0, possibly by the emission and absorption of phonons and/or photons. This exchange is absent in elastic resistors. The principle of energy conservation requires that the total energy entering the channel is zero

$$E_1 + E_2 + E_0 = 0. \tag{15.2}$$

This could be called an example of the *first law of thermodynamics*. However, there is yet another principle

$$\frac{E_1 - \mu_1 N_1}{T_1} + \frac{E_2 - \mu_2 N_2}{T_2} + \frac{E_0}{T_0} \leq 0 \tag{15.3}$$

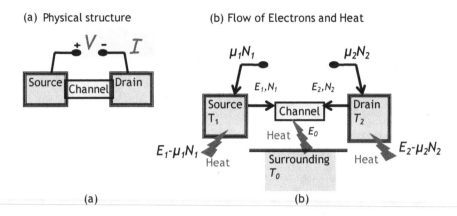

Fig. 15.1 (a) Physical structure (b) the flow of electrons and heat in (a) can be depicted in general terms as shown. For an elastic resistor, $E_0 = 0$.

known as the *second law of thermodynamics*. Unlike the first law, the second law involves an inequality. While most people are comfortable with the first law or the principle of energy conservation, the second law still continues to excite debate and controversy. And yet in some ways the second law embodies ideas that we know from experience. Suppose for example we assume all contacts to be at the same temperature ($T_2 = T_1 = T_0$). In this case Eq. (15.3) simply says that the total heat absorbed from the surroundings

$$(E_1 - \mu_1 N_1) + (E_2 - \mu_2 N_2) + E_0 \leq 0. \tag{15.4}$$

Making use of Eq. (15.2), this implies

$$\mu_1 N_1 + \mu_2 N_2 \geq 0. \tag{15.5}$$

The total energy exchanged in the process $E_1 + E_2 + E_0$ has two parts: One that came from the thermal energy of the surroundings and the other that came from the battery. Equation (15.4) tells us that the former must be negative, and Eq. (15.5) tells us that the latter must be positive. In other words, *we can take energy from a battery and dissipate it as heat, but we cannot take heat from the surroundings and charge up our battery if everything is at the same temperature.*

This should come as no surprise to anybody. After all if we could use heat from our surroundings to charge a battery (perhaps even run a car!) then there would be no energy problem. But the point to note is that this is not prohibited by the first law since energy would still be conserved. It is the second law that makes a distinction between the energy stored in a battery and the thermal energy in our surroundings.

The first is easily converted into the second, but not the other way around because thermal energy is distributed among many degrees of freedom. We can take energy from one degree of freedom and distribute it among many degrees of freedom, but we cannot take energy from many degrees of freedom and concentrate it all in one. This intuitive feeling is quantified and generalized by the second law (Eq. (15.3)) based on solid experimental evidence.

For example if we have multiple "contacts" at different temperatures then as we saw in Chapter 13 it is possible to take heat from the hotter contact, dump a part of it in the colder contact, use the difference to charge up a battery and still be compliance with the second law. Are all the things we have discussed so far in compliance with the second law?

The answer is yes. For the elastic resistor $E_0 = 0$, and we can write the second law from Eq. (15.3) in the form

$$\frac{\varepsilon - \mu_1}{T_1} N_1 + \frac{\varepsilon - \mu_2}{T_2} N_2 \leq 0$$

where we have written $E_1 = \varepsilon N_1$ and $E_2 = \varepsilon N_2$, assuming that each electron entering and exiting the channel has an energy of ε. Making use of Eq. (15.1) this means that

$$\left(\frac{\varepsilon - \mu_1}{T_1} - \frac{\varepsilon - \mu_2}{T_2} \right) N_1 \leq 0.$$

Our description of the elastic resistor always meets this condition, since the flow of electrons is determined by $f_1 - f_2$, as we saw in Chapter 3. N_1 is positive indicating electron flow from source to drain if $f_1(\varepsilon) > f_2(\varepsilon)$ that is, if

$$\frac{1}{1 + \exp\left(\dfrac{\varepsilon - \mu_1}{kT} \right)} > \frac{1}{1 + \exp\left(\dfrac{\varepsilon - \mu_2}{kT} \right)}$$

$$\frac{\varepsilon - \mu_1}{T_1} < \frac{\varepsilon - \mu_2}{T_2}.$$

Similarly we can show that N_1 is negative if

$$\frac{\varepsilon - \mu_1}{T_1} > \frac{\varepsilon - \mu_2}{T_2}.$$

In either case we have

$$\left(\frac{\varepsilon - \mu_1}{T_1} - \frac{\varepsilon - \mu_2}{T_2} \right) N_1 \leq 0$$

thus ensuring that the second law is satisfied. But what if we wish to go beyond the elastic resistor and include energy exchange within the channel. What would we need to ensure that we are complying with the second law?

15.2   Asymmetry of Absorption and Emission

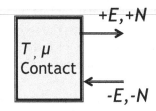

The answer is that our model needs to ensure that for all processes involving the exchange of electrons with a contact held at a potential μ and temperature T, the probability of *absorbing* E, N be related to the probability of *emitting* E, N by the relation

$$\frac{P(+E, +N)}{P(-E, -N)} = \exp \left(-\frac{E - \mu N}{kT} \right). \tag{15.6}$$

If only energy is exchanged, but not electrons, then the relation is modified to

$$\frac{P(+E)}{P(-E)} = e^{-E/kT}. \tag{15.7}$$

To see how this relation (Eq. (15.6)) ensures compliance with the second law (Eq. (15.3)), consider the process depicted in Fig. 15.1 involving energy and/or electron exchange with three different "contacts". Such a process should have a likelihood proportional to

$$P(E_1, N_1) P(E_2, N_2) P(E_0)$$

while the likelihood of the reverse process will be proportional to

$$P(-E_1, -N_1) P(-E_2, -N_2) P(-E_0).$$

In order for the former to dominate their ratio must exceed one:

$$\frac{P(+E_1,+N_1)P(+E_2,+N_2)P(+E_0)}{P(-E_1,-N_1)P(-E_2,-N_2)P(-E_0)} \geq 1.$$

If all processes obey the relations stated in Eqs. (15.6) and (15.7), we have

$$\exp\left(-\frac{E_1-\mu_1N_1}{T_1}\right)\exp\left(-\frac{E_2-\mu_2N_2}{T_2}\right)\exp\left(-\frac{E_0}{T_0}\right) \geq 1 \quad (15.8)$$

which leads to the second law stated in Eq. (15.3), noting that $\exp(-x)$ is greater than one, only if x is less than zero. Note that the equality in Eq .(15.3) corresponds to the forward probability being only infinitesimally larger than the reverse probability, implying a very slow net forward rate. To make the "reaction" progress faster, the forward probability needs to exceed the reverse probability significantly, corresponding to the inequality in Eq. (15.3).

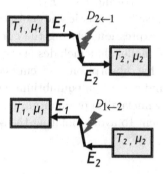

So how do we make sure our model meets the requirement in Eq. (15.6)? Consider for example a conductor with one inelastic scatterer in the middle separating a region having an energy level at E_1 from another having a level at E_2. Electrons flow from contact 1 to 2 by a process of emission whose probability is given by

$$D_{2\leftarrow1}f_1(E_1)(1-f_2(E_2))$$

while the flow from 2 to 1 requires an absorption process with probability

$$D_{1\leftarrow2}f_2(E_2)(1-f_1(E_1)).$$

Since one process involves emission while the other involves absorption, the rates should obey the requirement imposed by Eq. (15.7):

$$\frac{D_{2\leftarrow1}}{D_{1\leftarrow2}} = e^{(E_1-E_2)/kT_0} \qquad (15.9)$$

as we had stated earlier in Chapter 12 in a different context (Eq. (12.7)). T_0 is the temperature of the surroundings with which electrons exchange energy. The current in such an inelastic resistor would be given by an expression of the form (suppressing the arguments E_1 and E_2 for clarity)

$$I \sim D_{2\leftarrow1}f_1(1 - f_2) - D_{1\leftarrow2}f_2(1 - f_1) \qquad (15.10)$$

which reduces to the familiar form for elastic resistors

$$I \sim (f_1 - f_2)$$

only if

$$D_{2\leftarrow1} = D_{1\leftarrow2}$$

corresponding to elastic scattering $E_2 = E_1$.

Ordinary resistors have both elastic and inelastic scatterers intertwined and there is no simple expression relating the current to f_1 and f_2. The bottom line is that any model that includes energy exchange in the channel should make sure that absorption and emission rates are related by Eq. (15.9) if the surroundings are in equilibrium with a fixed temperature. Any transport theory, semiclassical or quantum needs to make sure it complies with this requirement to avoid violating the second law.

15.3 Entropy

Related video lecture available at course website, Unit 4: L4.6.

The asymmetry of emission and absorption expressed by Eq. (15.6) is actually quite familiar to everyone, indeed so familiar that we may not recognize it. We all know that if we take a hydrogen atom and place its lone electron in an excited (say 2p) state, it will promptly emit light and descend to the 1s state. But an electron placed in the 1s state will stay there forever. We justify it by saying that the electron "naturally" goes to its lowest energy state. But there is really nothing natural about this. Any mechanical interaction (quantum or classical) that takes an electron from 2p to 1s will also take it from 1s to 2p.

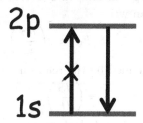

The natural descent of an electron to its lowest energy state is driven by a force that is not mechanical in nature. It is "entropic" in origin, as we will try to explain. Basically it comes from a *property of the surroundings* expressed by Eq. (15.6) which tells us that it is much harder to absorb anything from a reservoir, compared to emitting something into it. At zero temperature, a system can only emit and never absorb, and so an electron in state 2p can emit its way to the lowest energy state 1s, but an electron in state 1s can go nowhere.

This behavior is of course quite well-established and does not surprise anyone. But it embodies the key point that makes transport and especially quantum transport such a difficult subject in general. Any theoretical model has to include entropic processes in addition to the familiar mechanical forces. So where does the preferential tendency to lose energy rather than gain energy from any "reservoir" come from? Equation (15.6) can be understood by noting that when the electron loses energy the contact gains in energy so that the ratio of the rate of losing energy to the rate of gaining energy is equal to the ratio of the density of states at $E_0 + \epsilon$ to that at E_0 (Fig. 15.2):

$$\frac{P(-\varepsilon)}{P(+\varepsilon)} = \frac{W(E_0 + \varepsilon)}{W(E_0)}.$$

Here $W(E)$ represents the number of states available at an energy range E in the contact which is related to its entropy by the Boltzmann relation

$$S = k \ln(W) \tag{15.11}$$

so that

$$\frac{P(-\varepsilon)}{P(+\varepsilon)} = \exp\left(\frac{S(E_0 + \varepsilon) - S(E_0)}{k}\right). \tag{15.12}$$

Assuming that the energy exchanged ε is very small compared to that of the large contact, we can write

$$S(E_0 + \varepsilon) - S(E_0) \approx \varepsilon \left(\frac{dS}{dE}\right)_{E=E_0} = \frac{\varepsilon}{T}$$

with the temperature defined by the relation

$$\frac{1}{T} = \left(\frac{dS}{dE}\right)_{E=E_0}. \tag{15.13}$$

This is of course a very profound result saying that regardless of the detailed construction of any particular reservoir, as long as it is in equilibrium, dS/dE can be identified as its temperature.

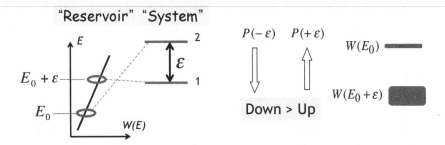

Fig. 15.2 Electrons preferentially go down in energy because it means more energy for the "reservoir" with a higher density of states. It is as if the lower state has a far greater "weight" as indicated in the lower panel.

If we accept this, then Eq. (15.11) gives us the basic relation that governs the exchange of energy with any "reservoir" in equilibrium with a temperature T:

$$\frac{P(-\varepsilon)}{P(+\varepsilon)} = e^{\varepsilon/kT}$$

as we stated earlier (see Eq. (15.7)).

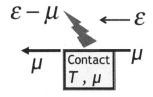

If the emission of energy involves the emission of an electron which eventually leaves the contact with an energy μ, then ε should be replaced by $\varepsilon - \mu$, as indicated in Eq. (15.6). The key idea is the same as what we introduced in Fig. 13.8 when discussing thermoelectric effects, namely that when an electron is added to a reservoir with energy ε, an amount $\varepsilon - \mu$ is dissipated as heat, the remaining μ representing an increase in the energy of the contact due to the added electron. Indeed that is the definition of the electrochemical potential μ. Eventually the added electron leaves the contact as shown.

15.3.1 *Total entropy increases continually*

Now that we have defined the concept of entropy, we can use it to restate the second law from Eq. (15.3). If we look at Fig. 15.1b we note that $E_1 - \mu_1 N_1$ represents the energy exchange with a "reservoir" at T_1, $E_2 - \mu_2 N_2$ represents the energy exchange with a "reservoir" at T_2, and E_0 represents the energy exchange with a "reservoir" at T_0.

Based on the definition of temperature in Eq. (15.13), we can write the corresponding changes in entropy ΔS_1, ΔS_2 and ΔS_0 as shown below

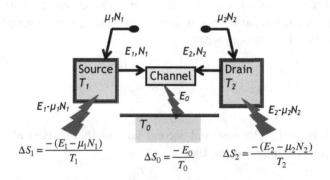

Note that these are exactly the same terms (except for the negative sign) appearing in Eq. (15.3), which can now be restated as

$$(\Delta S)_1 + (\Delta S)_2 + (\Delta S)_0 \geq 0. \qquad (15.14)$$

In other words, the second law requires the total change in entropy of all the reservoirs to be positive.

15.3.2 *Free energy decreases continually*

At zero temperature, any system in coming to equilibrium with its surroundings, goes to its state having the lowest energy. This is because a

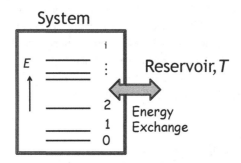

reservoir at zero temperature will only allow the system to give up energy, but not to absorb any energy. Interestingly, at non-zero temperatures, one can define a quantity called the *free energy*

$$F = E - TS \qquad (15.15)$$

such that at equilibrium a system goes to its state with minimum free energy. At $T = 0$, the free energy, F is the same as the total energy, E.

To see this, consider a system that can exchange energy with a reservoir such that the total energy is conserved. Using the subscript "R" for reservoir quantities we can write

$$\Delta E + \Delta E_R = 0 \qquad (15.16a)$$

$$\Delta S + \Delta S_R \geq 0 \qquad (15.16b)$$

which are basically the first and second laws of thermodynamics that we have been discussing. Noting that

$$\Delta S_R = \frac{\Delta E_R}{T}$$

we can combine Eqs. (15.16) to write

$$\Delta F \equiv \Delta E - T\Delta S \leq 0 \qquad (15.17)$$

which tells us that all energy exchange processes permitted by the first and second laws will cause the free energy to decrease, so that the final equilibrium state will be one with minimum free energy.

15.4 Law of Equilibrium

Related video lecture available at course website, Unit 4: L4.7.

The preferential tendency to lose energy rather than gain energy from any surrounding "reservoir" as expressed in Eq. (15.6) leads to a universal law stating that any system in equilibrium having states i with energy E_i and with N_i particles will occupy these states with probabilities

$$p_i = \frac{1}{Z} e^{-(E_i - \mu N_i)/kT} \tag{15.18}$$

where Z is a constant chosen to ensure that all the probabilities add up to one. To see this we note that all reservoirs in equilibrium have the property

$$\frac{P(+E, +N)}{P(-E, -N)} = \exp\left(-\frac{E - \mu N}{kT}\right). \tag{15.19}$$

$$E = E_1 - E_2,$$
$$N = N_1 - N_2$$

"Fock space"

Suppose we have a system with two states as shown exchanging energy and electrons with the surroundings. At equilibrium, we require upward transitions to balance downward transitions, so that

$$p_2 P(E, N) = p_1 P(-E, -N).$$

Making use of Eq. (15.6), we have

$$\frac{p_1}{p_2} = \frac{P(+E, +N)}{P(-E, -N)} = \exp\left(-\frac{(E_1 - \mu N_1) - (E_2 - \mu N_2)}{kT}\right).$$

It is straightforward to check that the probabilities given by Eq. (15.18) satisfy this requirement and hence represent an acceptable equilibrium solution. How can we have a law of equilibrium so general that it can be applied to all systems irrespective of its details? Because as we noted earlier *it comes from the property of the surroundings and not the system.*

Equation (15.18) represents the key principle or equilibrium statistical mechanics, Feynman (1965) called it the "summit". But it looks a little different from the two equilibrium distributions we introduced earlier, namely the Fermi function (Eq. (2.2)) and the Bose function (Eq. (14.5)).

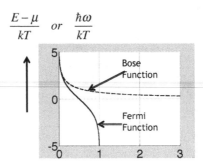

Fig. 15.3   The Fermi function (Eq. (2.2)) and the Bose function (Eq. (14.5)).

Figure 15.3 shows these two functions which look the same at high energies but deviate significantly at low energies. Electrons obey the exclusion principle and so the occupation $f(E)$ is restricted to values between 0 and 1. The Bose function is not limited between 0 and 1 since there is no exclusion principle.

Interestingly, however, both the Bose function and the Fermi function are special cases of the general law of equilibrium in Eq. (15.18). To see this, however, we need to introduce the concept of *Fock space* since the energy levels appearing in Eq. (15.18) do not represent the one-electron states we have been using throughout the book. They represent the so-called Fock space states, a new concept that needs some discussion.

15.5   Fock Space States

Consider a simple system with just one energy level, ε. In the one electron picture we think of electrons going in and out of this level. In the Fock space picture we think of the two possible states of the system, one corresponding to an empty state with energy $E = 0$, and one corresponding to a filled state with energy $E = \varepsilon$ as shown.

One-electron picture   "Fock space"

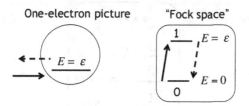

When an electron comes in the system goes from the empty state (0) to the full state (1), while if an electron leaves, the system goes from 1 to 0. Applying the general law of equilibrium (Eq. (15.18)) to the Fock space states, we have

$$p_0 = 1/Z \quad \text{and} \quad p_1 = \frac{e^{-x}}{Z}$$

$$\text{where } x \equiv (\varepsilon - \mu)/kT. \tag{15.20}$$

Since the two probabilities p_0 and p_1 must add up to one, we have

$$Z = 1 + e^{-x}$$

$$p_0 = \frac{1}{e^{-x} + 1} = 1 - f_0(\varepsilon)$$

$$p_1 = \frac{e^{-x}}{e^{-x} + 1} = \frac{1}{e^x + 1} = f_0(\varepsilon).$$

The probability of the system being in the full state, p_1 thus equals the Fermi function while the probability of the system being in the empty state, p_0 equals one minus the Fermi function, as we would expect.

15.5.1 *Bose function*

The Bose function too follows from Eq. (15.18), but we need to apply it to a system where the number of particles go from zero to infinity. Fock space states for electrons on the other hand are restricted to just zero or one because of the exclusion principle.

Equation (15.18) then gives us the probability of the system being in the N-photon state as

$$p_N = \frac{e^{-Nx}}{Z}, \quad \text{where } x \equiv \frac{\hbar\omega}{kT}.$$

To ensure that all probabilities add up to one, we have

$$Z = \sum_{N=0}^{\infty} e^{-Nx} = \frac{1}{1 - e^{-x}}$$

"Fock space"
for photons

so that the average number of photons is given by

$$n = \sum_{N=0}^{\infty} N p_N = \frac{1}{Z} \sum_{N=0}^{\infty} N e^{-Nx}.$$

Noting that

$$\sum_{N=0}^{\infty} N e^{-Nx} = -\frac{d}{dx} \sum_{N=0}^{\infty} e^{-Nx} = -\frac{d}{dx} Z$$

we can show with a little algebra that

$$n = \frac{1}{e^x - 1}$$

which is the Bose function stated earlier in Eq. (14.5).

The reason we have $E - \mu$ appearing in the Fermi function for electrons but not $\hbar\omega - \mu$ in the Bose function for photons or phonons is that the latter are not conserved. As we discussed in Section 15.3, when an electron enters the contact with energy E, it relaxes to an average energy of μ, and the energy dissipated is $E - \mu$. But when a photon or a phonon with energy $\hbar\omega$ is emitted or absorbed, the energy dissipated is just that. However, there are conserved particles (not photons or phonons) that also obey Bose statistics, and the corresponding Bose function has $E - \mu$ and not just E.

15.5.2 *Interacting electrons*

The general law of equilibrium (Eq. (15.18)) not only gives us the Fermi and Bose functions but in principle can also describe the equilibrium state of complicated interacting systems, if we are able to calculate the appropriate

Fock space energies. Suppose we have an interacting system with two one-electron levels corresponding to which we have four Fock space states as shown labeled 00, 01, 10 and 11. The 11 state with both levels occupied has an extra interaction energy U_0 as indicated.

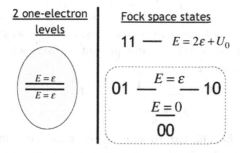

What is the average number of electrons if the system is in equilibrium with an electrochemical potential μ? Once again defining,

$$x \equiv \frac{\varepsilon - \mu}{kT}$$

we have from Eq. (15.18)

$$p_{00} = \frac{1}{Z}$$

$$p_{01} = p_{10} = \frac{e^{-x}}{Z}$$

$$p_{11} = \frac{e^{-2x}}{Z} e^{-U_0/kT}.$$

The average number of electrons is given by

$$n = 0 \cdot p_{00} + 1 \cdot p_{01} + 1 \cdot p_{10} + 2 \cdot p_{11}$$
$$= \frac{2(e^{-x} + e^{-2x} e^{-U_0/kT})}{Z}.$$

We could work out the details for arbitrary interaction energy U_0, but it is instructive to look at two limits. Firstly, the non-interacting limit with $U_0 \to 0$ for which

$$Z = 1 + 2e^{-x} + e^{-2x} = (1 + e^{-x})^2$$

so that with a little algebra we have

$$n = \frac{2}{1 + e^{(\varepsilon - \mu)/kT}}, \quad U_0 \to 0 \tag{15.21}$$

equal to the Fermi function times two as we might expect since there are two non-interacting states. The other limit is that of strongly interacting electrons for which $Z = 1 + 2e^{-x}$ so that

$$n = \frac{1}{1 + \frac{1}{2}e^{(\varepsilon-\mu)/kT}}, \quad U_0 \to \infty \tag{15.22}$$

a result that does not seem to follow in any simple way from the Fermi function. With g one-electron states present, it takes a little more work to show that the number is

$$n = \frac{1}{1 + \frac{1}{g}e^{(\varepsilon-\mu)/kT}}, \quad U_0 \to \infty. \tag{15.23}$$

This result may be familiar to some readers in the context of counting electrons occupying localized states in a semiconductor.

Equilibrium statistical mechanics is a vast subject and we are of course barely scratching the surface. My purpose here is simply to give a reader a feel for the concept of Fock space states and how they relate to the one electron states we have generally been talking about.

This is important because the general law of equilibrium (Eq. (15.18)) and the closely related concept of entropy (Eq. (15.11)) are both formulated in terms of Fock space states. We have just seen how the law of equilibrium can be translated into one-electron terms for non-interacting systems. Next let us see how one does the same for entropy.

15.6 Alternative Expression for Entropy

Related video lecture available at course website, Unit 4: L4.8.

$$S = Nk \, \ell n \, 2$$

Consider a system of independent localized spins, like magnetic impurities in the channel. At equilibrium, half the spins randomly point up and the other half point down. What is the associated entropy? Equation (15.11) defines the entropy S as $k \ln(W)$, W being the total number of Fock space states accessible to the system. In the present problem we could argue that each spin has two possible states (up or down) so that a collection of N spins has a total of 2^N states:

$$W = 2^N \to S = k \ln(W) = Nk \ln(2). \qquad (15.24)$$

This is correct, but there is an alternative expression that can be used whenever we have a system composed of a large number of identical independent systems, like the N spin collection we are considering:

$$S = -Nk \sum_i \tilde{p}_i \ln(\tilde{p}_i) \qquad (15.25)$$

where the $\tilde{p}_i$'s denote the probabilities of finding an individual system in its ith state. An individual spin, for example has a probability of $1/2$ for being in either an up or a down state, so that from Eq. (15.25) we obtain

$$S = -Nk \left\{ \frac{1}{2} \ln\left(\frac{1}{2}\right) + \frac{1}{2} \ln\left(\frac{1}{2}\right) \right\} = Nk \ln(2)$$

exactly the same answer as before (Eq. (15.24)).

Equation (15.25), however, is more versatile in the sense that we can use it easily even if the $\tilde{p}_i$'s happen to be say $1/4$ and $3/4$ rather than $1/2$ for each. Besides it is remarkably similar to the expression for the Shannon entropy associated with the information content of a message composed of a string of N symbols each of which can take on different values i with probability $\tilde{p}_i$. In the next chapter I will try to elaborate on this point further.

Let me end this chapter simply by indicating how this new expression for entropy given in Eq. (15.25) is obtained from our old one that we used in Eq. (15.24). This is described in standard texts on statistical mechanics (see for example, Dill and Bromberg (2003)).

15.6.1 *From Eq. (15.24) to Eq. (15.25)*

Consider a very large number N of identical systems each with energy levels E_i occupied according to probabilities $\tilde{p}_i$, such that the number of these systems in state i is given by

$$N_i = N\tilde{p}_i.$$

The total number of ways in which we can have a particular set of N_i should equal W, so that from standard combinatorial arguments we can write

$$W = \frac{N!}{N_1! N_2! \dots}.$$

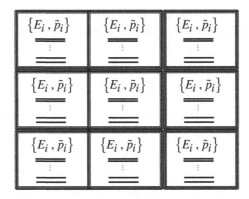

Taking the logarithm and using Stirling's approximation for large n

$$\ln(n!) \cong n \ln(n) - n$$

we have $\ln(W) = \ln(N!) - \ln(N_1!) - \ln(N_2!) - \ldots$

$$= N \ln(N) - N\tilde{p}_1 \ln(N\tilde{p}_1) - N\tilde{p}_2 \ln(N\tilde{p}_2) - \ldots$$

Making use of the condition that all the probabilities $\tilde{p}_i$ add up to one, we have

$$\ln(W) = -N(\tilde{p}_1 \ln(\tilde{p}_1) + \tilde{p}_2 \ln(\tilde{p}_2) + \ldots) = -N \sum_i \tilde{p}_i \ln(\tilde{p}_i).$$

This gives us W in terms of the probabilities, thus connecting the two expressions for entropy in Eq. (15.25) and Eq. (15.24).

15.6.2   *Equilibrium distribution from minimizing free energy*

One last observation before we move on. In general the system could be in some arbitrary state (not necessarily the equilibrium state) where each energy level E_i is occupied with some probability $\tilde{p}_i$. However, we have argued that for the equilibrium state, the probabilities $\tilde{p}_i$ are given by

$$[\tilde{p}_i]_{equilibrium} = \frac{1}{Z} e^{-E_i/kT} \equiv p_i \qquad (15.26)$$

where Z is a constant chosen to ensure that all the probabilities add up to one. We have also argued that the equilibrium state is characterized by a minimum in the free energy $F = E - TS$. Can we show that of all the possible choices for the probabilities $\tilde{p}_i$, the equilibrium distribution is p_i the one that will minimize the free energy?

Noting that the energy of an individual system is given by

$$E = \sum_i E_i \tilde{p}_i$$

and using S/N from Eq. (15.25) we can express the free energy as

$$F = \sum_i \tilde{p}_i (E_i + kT \ln(\tilde{p}_i)) \tag{15.27}$$

which can be minimized with respect to changes in $\tilde{p}_i$

$$dF = 0 = \sum_i d\tilde{p}_i (E_i + kT \ln(\tilde{p}_i) + kT)$$

$$= \sum_i d\tilde{p}_i (E_i + kT \ln(\tilde{p}_i))$$

noting that the sum of all probabilities is fixed, so that

$$\sum_i d\tilde{p}_i = 0.$$

We can now argue that in order to ensure that dF is zero for arbitrary choices of $d\tilde{p}_i$ we must have

$$E_i + kT \ln(\tilde{p}_i) = \text{a constant}$$

which gives us the equilibrium probabilities in Eq. (15.26).

Even if the system is not in equilibrium we can use Eq. (15.27) to calculate the free energy F of an out-of equilibrium system if we know the probabilities $\tilde{p}_i$. But the answer should be a number larger than the equilibrium value. In the next chapter, I will argue that in principle we can build a device that will harness the excess free energy

$$\Delta F = F - F_{eq}$$

of an out-of-equilibrium system to do useful work.

The excess free energy has two parts:

$$\underbrace{\Delta F}_{excess\ free\ energy} = \underbrace{\Delta E}_{excess\ energy} - T \times \underbrace{\Delta S}_{Information}. \tag{15.28}$$

The first part represents real energy, but the second represents information that is being traded to convert energy from the surrounding reservoirs into work. Let us now talk abut this "fuel value of information."

Chapter 16

Fuel Value of Information

Related video lecture available at course website, Unit 4: L4.9.

16.1 Introduction

A system in equilibrium contains no information, since the equilibrium state is independent of past history. Usually information is contained in systems that are stuck in some out of equilibrium state. We would like to argue that if we have such an out-of-equilibrium system, we can in principle construct a device that extracts an amount of energy less than or equal to

$$E_{available} = F - F_{eq} \tag{16.1}$$

where F is the free energy of the out-of-equilibrium system and F_{eq} is the free energy of the system once it is restored to its equilibrium state. Let me explain where this comes from.

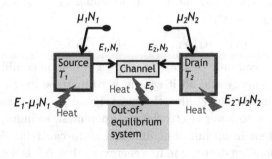

Fig. 16.1 An out-of-equilibrium system can in principle be used to construct a battery.

215

Consider the general scheme discussed in the last chapter, but with both contacts at the same temperature T and with the electrons interacting with some metastable system. Since this system is stuck in an out-of-equilibrium state we cannot in general talk about its temperature.

For example a collection of independent spins in equilibrium would be randomly half up and half down at any temperature. So if we put them into an all-up state, as shown below, we cannot talk about the temperature of this system. But we could still use Eq. (15.25) to find its entropy, which would be zero.

Equilibrium state, $S = Nk \, \ell n(2)$ *Out - of - equilibrium state,* $S = 0$

Fig. 16.2 A collection of N independent spins in equilibrium would be randomly half up and half down, but could be put into an out-of-equilibrium state with all spins pointing up.

With this in mind we could rewrite the second law by replacing

$$\frac{E_0}{T_0} \text{ with } - \Delta S$$

in Eq. (15.3):

$$\frac{E_1 - \mu_1 N_1}{T_1} + \frac{E_2 - \mu_2 N_2}{T_2} - \Delta S \leq 0. \tag{16.2}$$

Energy conservation requires that

$$E_1 + E_2 = -E_0 \equiv \Delta E \tag{16.3}$$

where ΔE is the change in the energy of the metastable system.

Combining Eqs. (16.2) and (16.3), assuming $T_1 = T_2 = T$, and making use of $N_1 + N_2 = 0$, we have

$$(\mu_1 - \mu_2)N_1 \quad \geq \quad \Delta E - T \Delta S = \Delta F. \tag{16.4}$$

Ordinarily, ΔF can only be positive, since a system in equilibrium is at its minimum free energy and all it can do is to increase its F. In that case, Eq. (16.4) requires that N_1 have the same sign as $\mu_1 - \mu_2$, that is, electrons flow from higher to lower electrochemical potential, as in any resistor.

But a system in an out-of-equilibrium state can relax to equilibrium with a corresponding decrease in free energy, so that ΔF is negative, and N_1 could have a sign opposite to that of $\mu_1 - \mu_2$, without violating Eq. (16.4).

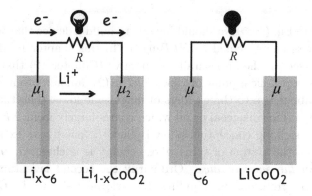

Electrons could then flow from lower to higher electrochemical potential, as they do inside a battery. The key point is that *a metastable non-equilibrium state can at least in principle be used to construct a battery.*

In a way this is not too different from the way real batteries work. Take the lithium ion battery for example. A charged battery is in a metastable state with excess lithium ions intercalated in a carbon matrix at one electrode. As lithium ions migrate out of the carbon electrode, electrons flow in the external circuit till the battery is discharged and the electrodes have reached the proper equilibrium state with the lowest free energy. The maximum energy that can be extracted is the change in the free energy. Usually the change in the free energy F comes largely from the change in the real energy E (recall that $F = E - TS$).

That does not sound too surprising. If a system starts out with an energy E that is greater than its equilibrium energy E_0, then as it relaxes, it seems plausible that a cleverly designed device could capture the extra energy $E - E_0$ and deliver it as useful work. What makes it a little more subtle, is that the extracted energy could come from the change in entropy as well.

For example the system of localized spins shown in Fig. 16.2 in going from the all-up state to its equilibrium state suffers no change in the actual energy, assuming that the energy is the same whether a spin points up or down. In this case the entire decrease in free energy comes from the increase in entropy:

$$\Delta E = 0 \tag{16.5a}$$

$$\Delta S = Nk \ln(2) \tag{16.5b}$$

$$\Delta F = \Delta E - T\Delta S = -NkT \ln(2). \tag{16.5c}$$

According to Eq. (16.5) we should be able to build a device that will deliver an amount of energy equal to $NkT \ln(2)$. In this chapter I will describe a device based on the anti-parallel spin valve (Chapter 12) that does just that. From a practical point of view, $NkT \ln(2)$, amounts to about 2.5 kJ per mole, about two to three orders of magnitude lower than the available energy of real fuels like coal or oil which comes largely from ΔE.

But the striking conceptual point is that the energy we extract is not coming from the system of spins whose energy is unchanged. *The energy comes from the surroundings.* Ordinarily the second law stops us from taking energy from our surroundings to perform useful work. But the information contained in the non-equilibrium state in the form of "negative entropy" allows us to extract energy from the surroundings without violating the second law.

From this point of view we could use the relation $F = E - TS$ to split up the right hand side of Eq. (16.1) into an actual energy and an info-energy that can be extracted from the surroundings by making use of the information available to us in the form of a deficit in entropy S relative to the equilibrium value S_{eq}:

$$E_{available} = \underbrace{E - E_{eq}}_{Energy} + \underbrace{T(S_{eq} - S)}_{Info-Energy} . \qquad (16.6)$$

For a set of independent localized spins in the all-up state, the available energy is composed entirely of info-energy: there is no change in the actual energy.

16.2   Information-driven Battery

Let us see how we could design a device to extract the info-energy from a set of localized spins. Consider a perfect anti-parallel spin-valve device (Chapter 12) with a ferromagnetic source that only injects and extracts upspin electrons and a ferromagnetic drain that only injects and extracts downspin electrons from the channel (Fig. 16.3). These itinerant electrons interact with the localized spins through an exchange interaction of the form

$$u + D \iff U + d \qquad (16.7)$$

where u, d represent up and down channel electrons, while U, D represent up and down localized spins. Ordinarily this "reaction" would be going equally in either direction. But by starting the localized spins off in a state with $U \gg D$, we make the reaction go predominantly from right to left and the resulting excess itinerant electrons u are extracted by one contact while the deficiency in d electrons is compensated by electrons entering the other contact. After some time, there are equal numbers of localized U and D spins and the reaction goes in either direction and no further energy can be extracted.

But what is the maximum energy that can be extracted as the localized spins are restored from their all up state to the equilibrium state? The answer is $NkT \ln(2)$ equal to the change in the free energy of the localized spins as we have argued earlier.

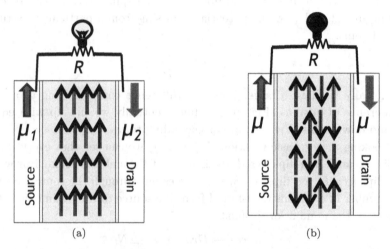

(a)   (b)

Fig. 16.3   An *info-battery*: (a) A perfect anti-parallel spin-valve device can be used to extract the excess free energy from a collection on N localized spins, all of which are initially up. (b) Eventually the battery runs down when the spins have been randomized.

But let us see how we can get this result from a direct analysis of the device. Assuming that the interaction is weak we expect the upspin channel electrons (u) to be in equilibrium with contact 1 and the downspin channel electrons (d) to be in equilibrium with contact 2, so that

$$f_u(E) = \frac{1}{\exp\left(\dfrac{E - \mu_1}{kT}\right) + 1}$$

$$\text{and} \quad f_d(E) = \frac{1}{\exp\left(\dfrac{E - \mu_2}{kT}\right) + 1}. \tag{16.8}$$

Assuming that the reaction

$$u + D \iff U + d$$

proceeds at a very slow pace so as to be nearly balanced, we can write

$$P_D\, f_u\, (1 - f_d) = P_U\, f_d\, (1 - f_u)$$

$$\frac{P_U}{P_D} = \frac{f_u}{1 - f_u}\, \frac{1 - f_d}{f_d} = e^{\Delta\mu/kT} \tag{16.9}$$

where $\quad \Delta\mu \equiv \mu_1 - \mu_2.$

Here we assumed a particular potential $\mu_{1,2}$ and calculated the corresponding distribution of up and down localized spins. But we can reverse this argument and view the potential as arising from a particular distribution of spins.

$$\Delta\mu = kT \ln\left(\frac{P_U}{P_D}\right). \tag{16.10}$$

Initially we have a larger potential difference corresponding to a preponderance of upspins (Fig. 16.3a), but eventually we end up with equal up and down spins (Fig. 16.3b) corresponding to $\mu_1 = \mu_2 = \mu$.

Looking at our basic reaction (Eq. (16.7)) we can see that everytime a D flips to an U, a u flips to a d which goes out through the drain. But when a U flips to a D, a d flips to a u which goes out through the source. So the net number of electrons transferred from the source to the drain equals the change in the number of U spins:

$$n(Source \to Drain) = -\Delta N_U$$

We can write the energy extracted as the potential difference times the number of electrons transferred

$$E = -\int_{Initial}^{Final} \Delta\mu\, dN_U. \tag{16.11}$$

Making use of Eq. (16.10) and noting that $N_U = N\, P_U$ we can write

$$E = -NkT \int_{Initial}^{Final} (\ln(P_U) - \ln(P_D))\, dP_U.$$

Noting that

$$dP_U + dP_D = 0$$

and that

$$S = -Nk(P_U \ln(P_U) + P_D \ln(P_D)) \tag{16.12}$$

we can use a little algebra to rewrite the integrand as

$$(\ln(P_U) - \ln(P_D))dP_U = d(P_U \ln(P_U) + P_D \ln(P_D)) = -\frac{dS}{Nk}$$

so that $\quad E = T \int_{Initial}^{Final} dS = T\Delta S$

which is the basic result we are trying to establish, namely that the metastable state of all upspins can in principle be used to construct a battery that could deliver upto

$$T\Delta S = NkT \ln(2)$$

of energy to an external load.

16.3   Fuel Value Comes from Knowledge

A key point that might bother a perceptive reader is the following. We said that the state with all spins up has a higher free energy than that for a random collection of spins: $F > F_0$, and that this excess free energy can in principle be extracted with a suitable device.

But what is it that makes the random collection different from the ordered collection. As Feynman put it, we all feel that it is unusual to see a car with a license plate # 9999. But it is really just as remarkable to see a car with any specific predetermined number say 1439. Similarly if we really knew the spins to be in a very specific but seemingly random configuration like the one sketched here, its entropy would be zero, just like the all up configuration. The possibility of extracting energy comes not from the all up nature of the initial state, *but from knowing exactly what state it is in.*

But how would we construct our conceptual battery to extract the energy from a random but known configuration? Consider a simple configuration that is not very random: the top half is up and the bottom half is down. Ordinarily this would not give us any open circuit voltage, the top half cancels the bottom half. But we could connect it as shown in Fig. 16.4 reversing the contacts for the left and right halves and extract energy.

Following the same principle we could construct a device to extract energy from a more random collection too. The key point is to know the exact configuration so that we can design the contacts accordingly.

Of course these devices would be much harder to build than the one we started with for the all-up configuration. But these devices are just intended to be conceptual constructs intended to illustrate a point. The

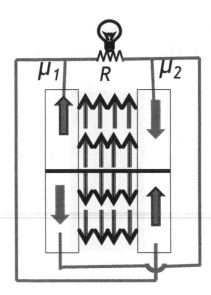

Fig. 16.4   A suitably designed device can extract energy from any known configuration of spins.

point is that information consists of a system being in an out of equilibrium state and our knowing exactly which state it is in. This information can *in principle* be used to create a battery and traded for energy.

In the field of information theory, Shannon introduced the word entropy as a measure of the information content of a signal composed of a string of symbols i that appear with probability p_i

$$H = -\sum_i p_i \ln(p_i). \qquad (16.13)$$

This expression looks like the thermodynamic entropy (see Eq. (15.25)) except for the Boltzmann constant and there are many arguments to this day about the connection between the two. One could argue that if we had a system with states i with equilibrium probabilities p_i, then $k \times H$ represents the entropy of an equilibrium system carrying no information. As soon as someone tells us which exact state it is in, the entropy becomes zero so that the entropy is lowered by $(Nk) \times H$ increasing its free energy by $(NkT) \times H$. In principle, at least we could construct a battery to extract this excess free energy $(NkT) \times H$.

16.4   Landauer's Principle

The idea that a known metastable state can be used to construct a battery can be connected to Landauer's principle which talks about the minimum energy needed to erase one bit of information. In our language, erasure consists of taking a system from an equilibrium state (F_{eq}) to a known standard state (F):

$$F_{eq} = -NkT\, \ell n(2) \quad \longrightarrow \quad \textit{Erasure} \quad \longrightarrow \quad F = 0$$

Is there a minimum energy needed to achieve this? We have just argued that once the spin is in the standard state we can construct a battery to extract $(F - F_{eq})$ from it. In a cyclic process we could spend E_{erase} to go from F_{eq} to F, and then construct a battery to extract $F - F_{eq}$ from it, so that the total energy spent over the cycle equals

$$E_{erase} - (F - F_{eq})$$

which must be greater than zero, or we would have a perpetual source of energy. Hence

$$E_{erase} \geq (F - F_{eq})$$

which in this case yields Landauer's principle:

$$E_{erase} \geq NkT \ln(2).$$

It seems to us, that erasure need not necessarily mean putting the spins in an all-up state. More generally it involves putting them in a known state, analogous to writing a complicated musical piece on a magnetic disk. Also, the minimum energy of erasure need not necessarily be dissipated. It often ends up getting dissipated only because it is impractical to build an info-battery to get it back. Fifty years ago Landauer asked deep questions that were ahead of his time. Today with the progress in nanoelectronics, the questions are becoming more and more relevant, and some of the answers at least seem fairly clear. Quantum mechanics, however, adds new features some of which are yet to be sorted out and are being actively debated at this time.

16.5 Maxwell's Demon

Our info-battery could be related to Maxwell's famous demon (see for example, Lex (2005)) who was conjectured to beat the second law by letting cold molecules (depicted gray) go from left to right and hot molecules (depicted black) go from right to left so that after some time the right hand side becomes cold and the left hand side becomes hot (Fig. 16.5).

To see the connection with our "info-battery" in Fig. 16.3 we could draw the following analogy:

$$
\begin{array}{rcl}
Hot\ molecules & \leftrightarrow & up-spin\ electrons\ (u) \\
Cold\ molecules & \leftrightarrow & down-spin\ electrons\ (d) \\
Demon & \leftrightarrow & collection\ of\ localized\ spins\ (U,\ D) \\
Left\ of\ box & \leftrightarrow & source\ contact \\
Right\ of\ box & \leftrightarrow & drain\ contact.
\end{array}
$$

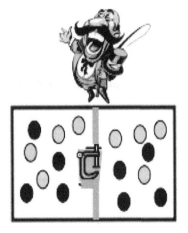

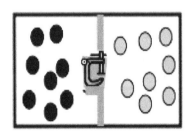

Fig. 16.5 Maxwell's demon creates a temperature difference by letting cold molecules go preferentially to the right.

Our battery is run by a set of all up localized spins that flip electrons up and send them to the source, while replacing the down-spin from the drain. The demon sends hot molecules to the left and cold molecules to the right, which is not exactly the same process, but similar.

The key point, however, is that the demon is making use of information rather than energy to create a temperature difference just as our info-battery uses the low entropy sate of the localized spins to create a potential difference. Like our localized spins, the demon too must gradually transition into a high entropy state that will stop it from discriminating between hot and cold molecules. Or as Feynman (1963) put it in one of his Lectures,

" .. *if we build a finite-sized demon, the demon himself gets so warm, he cannot see very well after a while.*"

Like our info-battery (Fig. 16.3), eventually the demon stops functioning when the entropy reaches its equilibrium value and all initial information has been lost. We started in Chapter 1 by noting how transport processes combine two very different types of processes, one that is force-driven and another that is entropy-driven. In these last two chapters, my objective has been to give readers a feeling for the concept of an "entropic force" that drives many everyday phenomena.

$$S = 0 \qquad\qquad S = Nk \ln(2)$$

The fully polarized state with $S = 0$ spontaneously goes to the unpolarized state with $S = Nk \ln(2)$, but to make it go the other way we need to connect a battery and do work on it. This directed flow physically arises from the fact that the fully polarized state represents a single state while the unpolarized state represents numerous (2^N) possibilities. It is this sheer number that drives the impurities spontaneously from the low entropy to the high entropy state and not the other way. Many real life phenomena are driven by such entropic forces which are very different from ordinary forces that take a system from a single state to another single state.

Interestingly these deep concepts are now finding applications in the design of *Boltzmann machines* that are having a major impact on the very active field of machine learning. The Boltzmann law tells us how to calculate the probability of different Fock space states (Section 15.5) in large systems. Today people are figuring out how to solve important problems by designing the interactions in large systems such that the highest probability state in Fock space embeds the correct solution to the problem. But that is a different story and this volume is already too long!

The main message I want to convey is that what makes the topic of transport so complicated is the intertwining of force-driven and entropy-driven phenomena. We have seen how the elastic resistor allows us to separate the two, with entropic processes confined to the contacts, and the channel described by semi-classical mechanics. It is time to move on to Part B to look at the quantum version of the problem with the channel described by quantum mechanics.

Suggested Reading

This book is based on a set of two online courses originally offered in 2012 on nanoHUB-U and more recently in 2015 on edX. These courses are now available in self-paced format at nanoHUB-U (https://nanohub.org/u) along with many other unique online courses.

In preparing the second edition we decided to split the book into parts A and B following the two online courses available on nanoHUB-U entitled *Fundamentals of Nanoelectronics*

Part A: Basic Concepts   Part B: Quantum Transport.

Video lecture of possible interest in this context: NEGF: A Different Perspective.

A detailed list of *video lectures* available at the course website corresponding to different sections of this volume (Part A: Basic Concepts) have been listed at the beginning.

Even this Second Edition represents lecture notes in unfinished form. I plan to keep posting additions/corrections at the book website.

This book is intended to be accessible to anyone in any branch of science or engineering, although we have discussed advanced concepts that should be of interest even to specialists, who are encouraged to look at my earlier books for additional technical details.

Datta S. (1995). Electronic Transport in Mesoscopic Systems
Datta S. (2005). Quantum Transport: Atom to Transistor

Cambridge University Press

Over 50 years ago David Pines in his preface to the Frontiers in Physics lecture note series articulated the need for both a consistent account of a field and the presentation of a definite point of view concerning it. That is what we have tried to provide in this book, with no intent to slight any other point of view or perspective.

The viewpoint presented here is unique, but not the topics we discuss. Each topic has its own associated literature that we cannot do justice to. What follows is a *very incomplete list* representing a small subset of the relevant literature, consisting largely of references that came up in the text.

Chapter 1

Figure 1.5 is reproduced from

McLennan M. *et al.* (1991) Voltage Drop in Mesoscopic Systems, Phys. Rev. B, 43, 13846

A recent example of experimental measurement of potential drop across nanoscale defects

Willke P. *et al.* (2015) Spatial Extent of a Landauer Residual-resistivity Dipole in Graphene Quantified by Scanning Tunnelling Potentiometry, Nature Communications, 6, 6399.

The transmission line model referenced in Section 1.8 is discussed in Section 9.4 and is based on

Salahuddin S. *et al.* (2005) Transport Effects on Signal Propagation in Quantum Wires, IEEE Trans. Electron Dev. 52, 1734

Some of the classic references on the non-equilibrium Green's function (NEGF) method

Martin P.C. and Schwinger J. (1959) Theory of Many-particle Systems I, Phys. Rev. 115, 1342

Kadanoff L.P. and Baym G. (1962) Quantum Statistical Mechanics, Frontiers in Physics, Lecture note series, Benjamin/Cummings

Keldysh (1965) Diagram Technique for Non-equilibrium Processes, Sov. Phys. JETP 20, 1018

The quote on the importance of the "channel" concept in Section 1.8 is taken from

Anderson P.W. (2010) 50 years of Anderson Localization, ed. E. Abrahams, Chapter 1, Thoughts on localization

The quote on the importance of physical pictures, even if approximate, in Section 1.9 is taken from

Feynman R.P. (1963) Lectures on Physics, vol. II-2, Addison-Wesley.

Part I: What Determines the Resistance

Chapter 4
For an excellent physical discussion of diffusion processes and the diffusion time (Eq. 4.7)

Berg H.C. (1993) Random Walks in Biology, Princeton University Press.

Two experiments reporting the discovery of quantized conductance in ballistic conductors.

van Wees, B.J. *et al.* (1988) Quantized Conductance of Points Contacts in a Two-Dimensional Electron Gas, Phys. Rev. Lett. 60, 848.
Wharam, D.A. *et al.* (1988) One-Dimensional Transport and the Quantisation of the Ballistic Resistance, J. Phys. C. 21, L209.

An experiment showing approximate conductance quantization in a hydrogen molecule

Smit R.H.M. *et al.* (2002) Measurement of the Conductance of a Hydrogen Molecule, Nature 419, 906.

Chapter 5
For an introduction to diagrammatic methods for conductivity calculation based on the Kubo formula, the reader could look at

Doniach S. and Sondheimer E.H. (1974), Green's Functions for Solid State Physicists, Frontiers in Physics Lecture Note Series, Benjamin/Cummings

Also Section 5.5 of Datta (1995) cited at the beginning.

There is an extensive body of work on subtle correlation effects in elastic resistors some of which have been experimentally observed. See for example,

Splettstoesser J. *et al.* (2010) Two-particle Aharonov-Bohm effect in Electronic Interferometers, Journal of Physics A: Mathematical and Theoretical 43, 354027.

Part II: Simple Model for Density of States

Chapter 6
A seminal paper in the field

Thouless, D. (1977). Maximum Metallic Resistance in Thin Wires, Phys. Rev. Lett. 39, 1167

An example of experiments showing electron density dependence of graphene conductance

Bolotin K.I. *et al.* (2008) Temperature Dependent Transport in Suspended Graphene, Phys. Rev. Lett. 101, 096802

A recent book with a thorough discussion of graphene-related materials

Torres L.E.F. Foa *et al.* (2014) Introduction to Graphene-based Nanomaterials, Cambridge University Press

Chapter 7
This discussion is based on a model pioneered by Mark Lundstrom that is widely used in the field.

Rahman A. *et al.* (2003) Theory of Ballistic Transistors, IEEE Trans. Electron Dev. 50, 1853 and references therein.
Lundstrom M.S. and Antoniadis D. (2014) Compact Models and the Physics of Nanoscale FETs, IEEE Trans. Electron Dev. 61, 225 and references therein.

Part III: What and Where is the Voltage Drop

Section 9.4 is based on Salahuddin S. et al. (2005) cited in Chapter 1.

Chapter 10 is based on Chapters 2–3 of Datta (1995) cited in the beginning.

A couple of papers by the primary contributors to the subject of discussion:

Büttiker M. (1988) Symmetry of Electrical Conduction, IBM J. Res. Dev.
32, 317
Imry Y and Landauer R (1999) Conductance Viewed as Transmission, Rev.
Mod. Phys. 71, S306

To learn more about the Onsager relations the reader could look at a book like

Yourgrau W., van der Merwe A., Raw G. (1982) Treatise on Irreversible
and Statistical Thermophysics, Dover Publications

Chapter 11
The paper that reported the first observation of the amazing quantization of the Hall resistance:

von Klitzing K. *et al.* (1980) New Method for High-Accuracy Determi-
nation of the Fine Structure Constant Based on Quantized Hall
Resistance, Phys. Rev. Lett. 45, 494.

For more on edge states in the quantum Hall regime the reader could look at Chapter 4 of Datta (1995) and references therein.

Chapter 12
An article on spin injection by one of the inventors of the spin valve

Fert A. *et al.* (2007) Semiconductors between Spin-Polarized Sources and
Drains, IEEE Trans. Electron Devices 54, 921

An article reviewing spin injection in semiconductors, the problems and solutions

Schmidt G. (2005) Concepts for Spin Injection into Semiconductors a Re-
view J. Phys. D: Appl. Phys. 38, R107

To learn more about our viewpoint on electrochemical potentials in materials with spin-momentum locking that is presented here, interested readers could look at

Sayed S., Hong S. and Datta S. (2016) Multi-Terminal Spin Valve on Channels with Spin-Momentum Locking, Scientific Reports 6, 35658 and references therein.

Part IV: Heat and Electricity

Chapter 13
On the use of the Landauer approach for thermoelectric effects

Butcher P.N. (1990) Thermal and Electrical Transport Formalism for Electronic Microstructures with Many Terminals, J. Phys. Condens. Matt. 2, 4869 and references therein.

On the thermoelectric effects in molecules

Baheti K. *et al.* (2008) Probing the Chemistry of Molecular Heterojunctions Using Thermoelectricity, Nano Letters 8, 715.
Paulsson M. and Datta S. (2003) Thermoelectric Effect in Molecular Electronics, Phys. Rev. B 67, 241403(R)

Chapter 14
This discussion draws on my collaborative work with Mark Lundstrom and Changwook Jeong.

Jeong C. *et al.* (2011) Full dispersion versus Debye model evaluation of lattice thermal conductivity with a Landauer approach, J. Appl. Phys. 109, 073718-8 and references therein.

A couple of other references on the subject

Majumdar A. (1993) Microscale Heat Conduction in Dielectric Thin Films Journal of Heat Transfer 115, 7
Mingo N. (2003) Calculation of Si nanowire thermal conductivity using complete phonon dispersion relations, Phys. Rev. B 68, 113308

Chapter 15
To learn more about these deep concepts, the reader could look at the many excellent texts on equilibrium statistical mechanics, such as

Dill K. and Bromberg S. (2003) Molecular Driving Forces, Statistical Thermodynamics in Chemistry and Biology, Garland Science
Kittel C. and Kroemer H. (1980) Thermal Physics, Freeman

Feynman R.P. (1965) Statistical Mechanics, Frontiers in Physics Lecture Note Series, Benjamin/Cummings

Chapter 16
The info-battery described here follows the discussion in

Datta S. (2008) Chapter 7: Nanodevices and Maxwell's demon in Nano Science and Technology eds. Tang Z., Sheng P., also available at https://arxiv.org/abs/0704.1623
See also Strasberg P. *et al.* (2014) Second laws for an information driven current through a spin valve, Phys. Rev. E 90, 062107

For more on Maxwell's demon and related issues

Feynman R.P. (1963) Lectures on Physics, vol. I-46, Addison-Wesley.
Leff H.S. and Rex A.F. (2003), Maxwell's Demon 2, IOP Publishing.

PART 5
Appendices

Appendix A

Derivatives of Fermi and Bose Functions

A.1 Fermi Function

$$f(x) \equiv \frac{1}{e^x + 1} \ , \ x \equiv \frac{E - \mu}{kT} \tag{A.1}$$

$$\frac{\partial f}{\partial E} = \frac{\partial f}{\partial x}\frac{\partial x}{\partial E} = \frac{\partial f}{\partial x}\frac{1}{kT} \tag{A.2a}$$

$$\frac{\partial f}{\partial \mu} = \frac{\partial f}{\partial x}\frac{\partial x}{\partial \mu} = -\frac{\partial f}{\partial x}\frac{1}{kT} \tag{A.2b}$$

$$\frac{\partial f}{\partial T} = \frac{\partial f}{\partial x}\frac{\partial x}{\partial T} = -\frac{\partial f}{\partial x}\frac{E - \mu}{kT^2}. \tag{A.2c}$$

From Eq. (A.2a) to Eq. (A.2c),

$$\frac{\partial f}{\partial \mu} = -\frac{\partial f}{\partial E} \tag{A.3a}$$

$$\frac{\partial f}{\partial T} = -\frac{E - \mu}{T}\frac{\partial f}{\partial E}. \tag{A.3b}$$

Equation (2.11) in Chapter 2 is obtained from a Taylor series expansion of the Fermi function around the equilibrium point

$$f(E, \mu) \approx f(E, \mu_0) + \left(\frac{\partial f}{\partial \mu}\right)_{\mu = \mu_0} (\mu - \mu_0).$$

From Eq. (A.3a),

$$\left(\frac{\partial f}{\partial \mu}\right)_{\mu=\mu_0} = \left(-\frac{\partial f}{\partial E}\right)_{\mu=\mu_0}.$$

Letting $f(E)$ stand for $f(E, \mu)$, and $f_0(E)$ stand for $f(E, \mu_0)$, we can write

$$f(E) = f_0(E) + \left(-\frac{\partial f}{\partial E}\right)_{\mu=\mu_0} (\mu - \mu_0)$$

which is the same as Eq. (2.11).

A.2 Bose Function

$$n(x) \equiv \frac{1}{e^x - 1}, \; x \equiv \frac{\hbar \omega}{kT} \tag{A.4}$$

$$\frac{\partial n}{\partial T} = \frac{dn}{dx} \frac{\partial x}{\partial T} = -\frac{\hbar \omega}{kT^2} \frac{dn}{dx}$$

$$\hbar \omega \frac{\partial n}{\partial T} = -kx^2 \frac{dn}{dx} = \frac{kx^2 e^x}{(e^x - 1)^2}. \tag{A.5}$$

Appendix B

Angular Averaging

This appendix provides the algebraic details needed to arrive at Eqs. (4.8) to (4.10) starting from

$$\bar{u} = \langle |v_z| \rangle, \quad \overline{D} = \langle v_z^2 \tau \rangle, \quad \lambda = \frac{2\overline{D}}{\bar{u}}.$$

B.1 One Dimension

$$\bar{u} = v, \quad \overline{D} = v^2 \tau.$$

1D

$-v$ $+v$

B.2 Two Dimensions

$$\bar{u} = \frac{\int\limits_{-\pi}^{+\pi} d\theta |v \cos\theta|}{\int\limits_{-\pi}^{\pi} d\theta} = v \frac{\int\limits_{-\pi/2}^{+\pi/2} d\theta \cos\theta}{\int\limits_{-\pi/2}^{\pi/2} d\theta} = \frac{2v}{\pi}$$

2D

$$\overline{D} = v^2 \tau \frac{\int\limits_0^{2\pi} d\theta \cos^2\theta}{\int\limits_0^{2\pi} d\theta} = \frac{v^2 \tau}{2}.$$

239

B.3   Three Dimensions

$$\bar{u} = \frac{\int\limits_{0}^{2\pi} d\phi \int\limits_{0}^{\pi} d\theta \sin\theta |v\cos\theta|}{\int\limits_{0}^{2\pi} d\phi \int\limits_{0}^{\pi} d\theta \sin\theta} = v \frac{\int\limits_{0}^{2\pi} d\phi \int\limits_{0}^{\pi/2} d\theta \sin\theta \cos\theta}{\int\limits_{0}^{2\pi} d\phi \int\limits_{0}^{\pi/2} d\theta \sin\theta} = \frac{v}{2}$$

$$\overline{D} = v^2\tau \frac{\int\limits_{0}^{2\pi} d\phi \int\limits_{0}^{\pi/2} d\theta \sin\theta \cos^2\theta}{\int\limits_{0}^{2\pi} d\phi \int\limits_{0}^{\pi/2} d\theta \sin\theta} = \frac{v^2\tau}{3}.$$

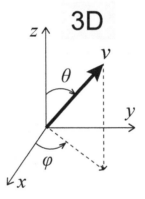

B.4   Summary

$$\bar{u} = \langle |v_z| \rangle = v \left\{ \overbrace{1}^{1D}, \overbrace{\frac{2}{\pi}}^{2D}, \overbrace{\frac{1}{2}}^{3D} \right\}$$

$$\overline{D} = \langle v_z^2 \tau \rangle = v^2\tau \left\{ \overbrace{1}^{1D}, \overbrace{\frac{1}{2}}^{2D}, \overbrace{\frac{1}{3}}^{3D} \right\}$$

$$\lambda \equiv \frac{2\overline{D}}{\bar{u}} = v\tau \left\{ \overbrace{2}^{1D}, \overbrace{\frac{\pi}{2}}^{2D}, \overbrace{\frac{4}{3}}^{3D} \right\}.$$

Appendix C

Current at High Bias for Non-degenerate Resistors

We saw in Chapter 7 that charging effects can be included into our current equation for elastic resistors by replacing $G(E)$ with $G(E - U)$:

$$I = \frac{1}{q} \int_{-\infty}^{+\infty} dE \, G(E - U) \, (f_1(E) - f_2(E)) \rightarrow \text{same as Eq. (7.1)}$$

where $G(E) = \dfrac{\sigma A}{L + \lambda} \rightarrow \dfrac{\sigma A}{L}$ *if we exclude interface resistance.*

In Section 7.5 we argued that the presence of an electric field in the channel causes an effective widening of the channel from the source to the drain that should cause an increase in the conductance (see Fig. 7.13) which can be included in a full numerical model (Fig. 7.14).

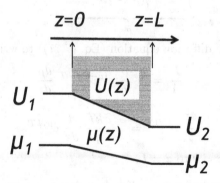

Interestingly, if we assume that the electrochemical potential is outside the band so that the Boltzmann approximation can be used for the Fermi function

$$f(E) \approx e^{-(E - \mu)/kT}$$

then we will show shortly that Eq. (7.1) can be used with a little modification

$$I = \frac{1}{q} \int_{-\infty}^{+\infty} dE \, G(E - U_0) \, (f_1(E) - f_2(E)) \qquad \text{(C.1)}$$

$$G(E) = \frac{\sigma(E) \, A}{l}, \text{ where } l \equiv \int_0^L dz \, e^{(U(z) - U_0)/kT}.$$

Unlike the point channel model, $U(z)$ is spatially varying and we use its maximum value as the reference potential U_0. Consequently

$$U(z) - U_0 \leq 0$$

$$\text{so that } \quad l \leq L \rightarrow \frac{\sigma A}{l} \geq \frac{\sigma A}{L}.$$

This increase in conductance reflects the effective increase in the channel width from source to drain that we discussed earlier (Fig. 7.13).

Derivation of Eq. (C.1):

This is a very nice result, but it can be derived only if we assume non-degenerate Boltzmann statistics to express the conductivity from Eq. (6.6) in the form

$$\sigma = \underbrace{\int_{-\infty}^{+\infty} \frac{dE}{kT} \, \sigma(E) \, e^{-E/kT}}_{\tilde{\sigma}} \, e^{(\mu - U)/kT}$$

and using it in the diffusion equation (Eq. (7.17)) to write

$$\frac{I}{A} = -\frac{\tilde{\sigma}}{q} \, e^{(\mu - U)/kT} \, \frac{d\mu}{dz}$$

$$\frac{I}{A} \, e^{U/kT} = -\frac{\tilde{\sigma} kT}{q} \, \frac{d}{dz} \, e^{\mu/kT}.$$

Integrating from $z = 0$ to $z = L$,

$$\frac{I}{A} \underbrace{\int_0^L dz \, e^{(U - U_0)/kT}}_{l} = \frac{\tilde{\sigma} \, k \, T}{q} \, e^{-U_0/kT} \left(e^{\mu_1/kT} - e^{\mu_2/kT} \right)$$

$$I = \frac{1}{q} \frac{A}{l} \, e^{-U_0/kT} \left(e^{\mu_1/kT} - e^{\mu_2/kT} \right) kT \int_{-\infty}^{+\infty} \frac{dE}{kT} \, \sigma(E) \, e^{-E/kT}$$

$$= \frac{1}{q} \int_{-\infty}^{+\infty} dE \, \frac{\sigma(E) \, A}{l} \, (f_1(E + U_0) - f_2(E + U_0))$$

$$= \frac{1}{q} \int_{-\infty}^{+\infty} dE \, \frac{\sigma(E - U_0) \, A}{l} \, (f_1(E) - f_2(E))$$

thus proving Eq. (C.1).

Equation (C.1) tells us that we can write the conductance of a section of length L as being that of a section of length l that is shorter than the physical length L. To use this result we need which requires us to know the potential profile $U(z)$ and this in turn would generally require a numerical solution.

But it is interesting that for non-degenerate conductors it is possible to write the current in a form that looks just like that for elastic resistors. For elastic conductors, however, the Boltzmann approximation is not needed, the current expression is valid for both degenerate and non-degenerate conductors.

Appendix D

Semiclassical Dynamics

$$E(\mathbf{x}, \mathbf{p}) = \frac{(\mathbf{p} - q\mathbf{A}) \cdot (\mathbf{p} - q\mathbf{A})}{2m} + U(\mathbf{x}) \tag{D.1}$$

where $q\mathbf{F} = -\nabla U$, Electric Field, $\mathbf{F}$
and $\mathbf{B} = \nabla \times \mathbf{A}$, Magnetic Field, $\mathbf{B}$

D.1 Semiclassical Laws of Motion

$$\mathbf{v} \equiv \frac{d\mathbf{x}}{dt} = \nabla_p E \tag{D.2a}$$

$$\frac{d\mathbf{p}}{dt} = -\nabla E \tag{D.2b}$$

where the gradient operators are defined as

$$\nabla E \equiv \hat{\mathbf{x}} \frac{\partial E}{\partial x} + \hat{\mathbf{y}} \frac{\partial E}{\partial y} + \hat{\mathbf{z}} \frac{\partial E}{\partial z}$$

$$\nabla_p E \equiv \hat{\mathbf{x}} \frac{\partial E}{\partial p_x} + \hat{\mathbf{y}} \frac{\partial E}{\partial p_y} + \hat{\mathbf{z}} \frac{\partial E}{\partial p_z}.$$

From Eqs. (D.1) and (D.2), we can show that

$$\mathbf{v} = (\mathbf{p} - q\mathbf{A})/m \tag{D.3}$$

$$\frac{d(\mathbf{p} - q\mathbf{A})}{dt} = q\left(\mathbf{F} + \mathbf{v} \times \mathbf{B}\right). \tag{D.4}$$

D.1.1   *Proof*

$$E(\mathbf{x}, \mathbf{p}) = \sum_j \frac{(p_j - qA_j(\mathbf{x}))^2}{2m} + U(\mathbf{x})$$

$$\longrightarrow v_i = \frac{\partial E}{\partial p_i} = \frac{(p_i - qA_i(\mathbf{x}))}{m} \longrightarrow \mathbf{v} = \frac{(\mathbf{p} - q\mathbf{A}(\mathbf{x}))}{m}$$

$$\frac{dp_i}{dt} = -\frac{\partial E}{\partial x_i} = -\frac{\partial U}{\partial x_i} + q\sum_j v_j \frac{\partial A_j}{\partial x_i}$$

$$\frac{d(p_i - qA_i(\mathbf{x}))}{dt} = -\frac{\partial U}{\partial x_i} + q\sum_j v_j \left(\frac{\partial A_j}{\partial x_i} - \frac{\partial A_i}{\partial x_j}\right)$$

$$= -\frac{\partial U}{\partial x_i} + q\sum_{j,n} v_j \varepsilon_{ijn} \left(\nabla \times \mathbf{A}\right)_n$$

$$\longrightarrow \frac{d(\mathbf{p} - q\mathbf{A})}{dt} = q\left(\mathbf{F} + \mathbf{v} \times \mathbf{B}\right).$$

In Chapter 9 we used the 1D version of Eqs. (D.2) to obtain the BTE (Eq. (9.6)). If we use the full 3D version we obtain

$$\frac{\partial f}{\partial t} + \mathbf{v} \cdot \nabla f + \mathbf{F} \cdot \nabla f = S_{op} f. \tag{D.5}$$

Appendix E

Transmission Line Parameters from BTE

In this Appendix, I will try to outline the steps involved in getting from Eq. (9.19) to Eq. (9.20), that is

$$\text{from} \quad \frac{\partial \mu}{\partial t} + v_z \frac{\partial \mu}{\partial z} - \frac{\partial E}{\partial t} = -\frac{\mu - \overline{\mu}}{\tau} \tag{E.1}$$

$$\text{to} \quad \frac{\partial(\mu/q)}{\partial z} = -(L_K + L_M)\frac{\partial I}{\partial t} - \frac{I}{\sigma A} \tag{E.2a}$$

$$\frac{\partial(\mu/q)}{\partial t} = -\left(\frac{1}{C_Q} + \frac{1}{C_E}\right)\frac{\partial I}{\partial z}. \tag{E.2b}$$

First we separate Eq. (E.1) into two equations for μ^+ and μ^-,

$$\frac{\partial \mu^+}{\partial t} + v_z \frac{\partial \mu+}{\partial z} - \frac{\partial E^+}{\partial t} = -\frac{\mu^+ - \overline{\mu}}{\tau} \tag{E.3a}$$

$$\frac{\partial \mu^-}{\partial t} - v_z \frac{\partial \mu-}{\partial z} - \frac{\partial E^-}{\partial t} = -\frac{\mu^- - \overline{\mu}}{\tau}. \tag{E.3b}$$

Next we add and subtract Eqs. (E.3a) and (E.3b) and use the relations

$$I = \frac{qM}{h}(\mu^+ - \mu^-) \quad \text{and} \quad \overline{\mu} = \frac{\mu^+ + \mu^-}{2} \equiv \mu$$

$$\text{to obtain} \quad 2\frac{\partial \mu}{\partial t} + \frac{v_z}{qM/h}\frac{\partial I}{\partial z} - \frac{\partial}{\partial t}\left(E^+ + E^-\right) = 0$$

$$\text{and} \quad \frac{1}{qM/h}\frac{\partial I}{\partial t} + 2v_z\frac{\partial \mu}{\partial z} - \frac{\partial}{\partial t}\left(E^+ + E^-\right) = -\frac{I}{(qM/h)\tau}.$$

247

Rearranging

$$\frac{\partial(\mu/q)}{\partial t} = -\frac{1}{C_Q}\frac{\partial I}{\partial z} + \frac{1}{2q}\frac{\partial}{\partial t}\left(E^+ + E^-\right) \tag{E.4a}$$

$$\frac{\partial(\mu/q)}{\partial z} = -L_K\frac{\partial I}{\partial t} + \frac{1}{2qv_z}\frac{\partial}{\partial t}\left(E^+ + E^-\right) - RI \tag{E.4b}$$

where L_K, C_Q are the quantities defined in Eq. (9.21).

Let us now consider the terms involving $E^\pm$. Assuming that the fields associated with a transverse electromagnetic (TEM) wave, they can be expressed in terms of $U(z,t)$ along with a vector potential $A_z(z,t)$ pointing along z, for which the energy is given by (Appendix C)

$$E = \frac{(p_z - qA_z(t))^2}{2m} + U(z,t). \tag{E.5}$$

Noting that

$$v_z = \frac{\partial E}{\partial p_z} = \frac{p_z - qA_z}{m}$$

we can write

$$\frac{\partial E}{\partial t} = v_z\left(-q\frac{\partial A_z}{\partial t}\right) + \frac{\partial U}{\partial t}$$

so that

$$\frac{\partial}{\partial t}\left(E^+ + E^-\right) = 2\frac{\partial U}{\partial t}$$

and

$$\frac{\partial}{\partial t}\left(E^+ - E^-\right) = 2v_z\left(-q\frac{\partial A_z}{\partial t}\right)$$

Eqs. (E.4) can then be written as

$$\frac{\partial(\mu/q)}{\partial t} = -\frac{1}{C_Q}\frac{\partial I}{\partial z} + \frac{\partial(U/q)}{\partial t} \tag{E.6}$$

$$\frac{\partial(\mu/q)}{\partial z} = -L_K\frac{\partial I}{\partial t} - \frac{\partial A_z}{\partial t} - RI \tag{E.7}$$

which reduces to Eqs. (E.2) noting that

$$A_z = L_M I, \quad U/q = Q/C_E$$

and making use of the continuity equation

$$\frac{\partial Q}{\partial t} + \frac{\partial I}{\partial z} = 0.$$

Index

Printed in the United States
By Bookmasters

Printed in the United States
By Bookmasters